Without Death, There Is No Resurrection

Did Jesus Die?

Dr. Octavian Caius Obeada

Copyright © 2024 Dr. Octavian Caius Obeada

All rights reserved. No part of this work may be reproduced, stored in a retrieval system or transmitted in any form by any means, electronic, mechanical, photocopying, recording, or otherwise, without written permission from the author.

All Scripture quotations, unless otherwise indicated, are taken from the Holy Bible, **ESV**: Scripture quotations marked ESV are taken from the ESV® Bible (The Holy Bible, English Standard Version®). Copyright © 2001 by Crossway, a publishing ministry of Good News Publishers. Used by permission. All rights reserved. **KJV**: Scripture quotations marked KJV are taken from the King James Version. Public domain.

ISBN: 9798320390543

DEDICATION

This book is dedicated to our Lord and Savior, Jesus Christ.
I extend my heartfelt gratitude to my wife, Sabina, for her enduring patience.
My appreciation also goes to my apologetics teachers and mentors: William Lane Craig, J.P. Moreland, Scott Smith, and John Warwick Montgomery, for their guidance and wisdom. Finally, I must acknowledge Laurentiu Nica, my fellow apologist and companion of 20 years, with whom I have shared the Gospel with Romanians globally.

Contents

	Preface	1
	Introduction	3
1	**Historical Context of Jesus' Life**	15
	Overview of 1st Century Judea under Roman Rule	15
	Roman Occupation of Judea	16
	Jewish Society and Religion	17
	Economic and Social Conditions	20
	Resistance and Unrest	23
	The Role of Religion in Society	26
2	**The Crucifixion – Accounts and Analysis**	29
	Description of the Crucifixion Event Based on the Gospels	30
	Historical and Archeological Evidence of Roman Crucifixion Practices	40
3	**Theological Perspectives on Jesus' Death**	51
	Christian Viewpoints of the Significance of Jesus' Death	51
	Atonement and Redemption in Christian Theology	63
4	**Did Jesus Die? – Examining the Historical Evidence**	71
	Analysis of Historical Evidence for Jesus' Death	71
	The "Bibliographical Test"	72
	Internal Evidence for the Gospels	83
	External Evidence for the Gospel Narratives	93
5	**Did Jesus Die? – Examining the Medical Evidence**	105
	Jesus' Exertion on Via Dolorosa	117
6	**Medical Evidence for Trauma**	123
	Physical Effects: Breathing Difficulty	131
	Physical Effects: Blood Loss and Shock	133
	Physical Effects: Dehydration and Exposure	134
	Psychological and Emotional Trauma	136
7	**Probability of Jesus' Death**	137
	Cause of Death	140
8	**Confirmation of Jesus' Death**	149
	The Spear Wound	149
	Lack of Movement	155
	Absence of Breathing	159
	No Response to Stimuli	165
	Purpose of Breaking Legs in Crucifixion	171
9	**Limitation in Analyzing the Data**	177
	Medicine in Antiquity	177
	Diagnostic Techniques	180
	Can You Diagnose it Based on Descriptive Facts?	181
	Limitation of the Analysis	182

10	How Confident We Are Regarding the Cause of Jesus' Death?	197
	Crucifixion Process	*197*
	Medical Analysis	*199*
	The Role of Scourging	*200*
	Medical Hypotheses on Specific Causes of Death	*202*
	Limitations of Medical Diagnosis	*203*
	Medical Providers and Their Opinions	*205*
11	Conclusion	215
	Cited or Recommended Literature	219
	Index	227

Preface

In the waning hours of many an evening, beneath the expanse of a starlit sky, I often found solace in the company of three or four childhood friends. We would gather at an unassuming bench in front of our building, using it as our rendezvous point - a fixture that we found as familiar as it was worn, with many of its wooden planks missing, telling tales of many such gatherings. This bench faced the children's park, a place that, by night, transformed into a realm of quiet shadows and half-lit silhouettes, with only a handful of lights still bravely warding off the complete embrace of darkness. Most families had retreated into their homes by then, their attentions turned to nightly news or the closing rituals before sleep, leaving the outside world to us and the night.

Surrounded by the gentle hush of the darkened park, our conversations would inevitably drift upwards, both literally and metaphorically. As we gazed into the heavens, the stars above seemed to cast their gentle glow upon us, a luminous tapestry that felt almost within reach. It was against this backdrop of serene semi-darkness, punctuated by the vivid spectacle of the night sky, that we delved into profound discussions about the existence of God, the possibility of life beyond our own world, and the infinite mysteries of the universe. These moments, shared in the night's tranquility, were a testament to the boundless curiosity and camaraderie that defined our youthful explorations.

On the cusp of a significant personal milestone, I found myself enveloped in the warmth of friendship, eager to share the decision that was about to reshape my path. My lineage embraced Baptist traditions, with my roots deeply connected to the faith for four generations. The decision to embrace the teachings of Jesus and undergo baptism was born of deep personal conviction, not merely a rite of passage inherited from my forebears.

As the announcement settled among my friends, it sparked a whirlwind of discussions with inquiries that ranged from curious to confrontational. One friend, with a blend of curiosity and challenge, posed the question that hovered on many minds: "Why are you getting baptized? Is it because of your parents?" The question, though provocative, found me unguarded and ready with a response that I had long pondered.

I had often proclaimed, with the boldness of youth, a desire to forge my own path, distinct from that of my parents. The notion of baptism, in particular, was a decision I was adamant would not be swayed by tradition or expectation. My fervor in discussions about my faith and decisions was palpable, driven by a conviction that what I was undertaking was a reflection of my personal belief, not a mirrored image of parental influence.

That night, under the canopy of stars and the watchful eyes of my friends, became a crucible for my faith. I found myself articulating the reasons for my belief in God, inviting my friends to embark on their own journeys of questioning, seeking, and perhaps finding their own convictions about life's great mysteries and the divine. While many questions lingered unanswered in the air, my resolve regarding my faith was unshakable – a testament to my conviction in Jesus' teachings and the promise of eternal life. This moment, shared amongst friends, was not just a declaration of faith but a defining assertion of my identity and beliefs.

Dr. Octavian Caius Obeada

Introduction

"Without Death, There Is No Resurrection: Did Jesus Die?" embarks on a deep and sweeping exploration into one of the most significant inquiries at the heart of Christian faith: the historical and physical reality of Jesus of Nazareth's death by crucifixion. This investigation explores more than just academic curiosity; through a rigorous amalgamation of historical research, archaeological findings, theological insight, and medical analysis, the book offers a multidimensional approach to understanding an event that has shaped the contours of history and faith in profound ways.

Historical Research: The book delves into the annals of history, examining the political, social, and religious context of 1st-century Judea, where Jesus lived and ministered. It scrutinizes the historical evidence for Jesus' crucifixion, drawing on ancient texts, including the canonical Gospels and writings from Roman and Jewish historians of the era. This historical exploration seeks to place Jesus' death within the broader narrative of Roman execution practices, Jewish expectations of messianic figures, and the tumultuous socio-political environment of the time.

Archaeological Findings: complement the historical narrative by bringing archaeological insights to the discussion. Discoveries from 1st-century Judea, including burial practices and artifacts related to Roman crucifixion, enrich the understanding of the physical realities of Jesus' execution. These findings help to bridge the gap between the biblical text and the historical context, providing tangible evidence of the practices and conditions described in the Gospel accounts.

Theological Insight: The book delves into the theological dimensions of Jesus' death with depth and sensitivity. The book examines the interpretation of the crucifixion within Christian theology as the ultimate act of sacrifice, offering redemption from sin for humanity. The book delves into the doctrines of atonement and resurrection, considering how people understand Jesus' death and subsequent resurrection as fulfilling Old Testament prophecies and the foundation of the New Covenant between God and humankind.

Medical Analysis: A unique aspect of the book is its incorporation of medical analysis to examine the physical aspects of crucifixion. Drawing on current medical knowledge, the book offers insights into the physiological effects of scourging, crucifixion, and death by asphyxiation – conditions that Jesus likely endured. This medical perspective provides a visceral understanding of the suffering involved in crucifixion, lending a profound depth to the theological reflections on Jesus' sacrifice.

"Without Death, There Is No Resurrection: Did Jesus Die?" invites readers into a space where faith and reason, history and theology, science and spirituality intersect. It challenges both believers and skeptics to consider the evidence and implications of Jesus' death and resurrection. This comprehensive journey not only seeks to answer the question whether Jesus died by crucifixion but also to understand the significance of this event in the broader context of human history and divine salvation. Through this exploration, the book aims to deepen the reader's appreciation for the central event of Christian faith, encouraging a reflective engagement with the themes of sacrifice, redemption, and hope that resonate through the ages.

Chapter One: Historical Context of Jesus' Life
The foundational chapter of this exploration sets out to recreate the complex tapestry of 1st-century Judea, a period characterized by its rich cultural, social, and political intricacies under the shadow of Roman occupation. This era stands as a critical juncture in history, where various forces – political ambition, religious fervor, and social dynamics – intertwine, shaping the landscape in which Jesus' ministry unfolds.

Roman Occupation of Judea: This segment delves into the political environment of Judea, emphasizing the Roman Empire's dominance over the region. It explores the mechanisms of Roman governance, imposing taxes, and the military presence, illustrating the overarching control and influence Rome exerted on the daily lives of the Judean populace. The juxtaposition of Roman law and Jewish tradition creates a setting ripe with tension, setting the stage for conflict and unrest.

Jewish Society and Religion: In the heart of this chapter lies an exploration of the religious and social structures that defined Jewish life. The chapter highlights the religious and social structures that shaped Jewish life, revealing the diversity within Jewish society. It examines different groups, like the Pharisees, Sadducees, Essenes, and Zealots, each with their unique beliefs and expectations of the Messiah. This section also examines the role of the Temple in Jerusalem, not only as a religious center but also as a symbol of Jewish identity and resistance against Roman influence.

Economic and Social Conditions: We scrutinize the economic realities of the time to understand the daily struggles faced by the common people. The disparity between the wealthy elite and the impoverished masses, exacerbated by heavy taxation and economic exploitation, underscores the social tensions that pervade Jewish society. This economic hardship, coupled with social stratification, fuels discontent and sets the scene for Jesus' teachings on wealth, poverty, and social justice.

Resistance and Unrest: Here, the narrative shifts to the undercurrents of resistance against Roman rule. The chapter explores various forms of opposition, from passive resistance and the non-violent protest of Jesus' teachings to the more direct confrontations led by revolutionary groups. This resistance is not merely a backdrop but a dynamic element that interacts with the broader narrative of Jesus' life and ministry.

The Role of Religion in Society: Finally, this section emphasizes the central role of religion in shaping societal norms and individual identities. It examines how religious beliefs and practices permeated every aspect of life in Judea, influencing social interactions, political aspirations, and personal convictions. Considering the societal longing for deliverance from Roman oppression, the book explores the expectations of a Messiah, which are deeply rooted in Jewish prophetic tradition.

By painting a detailed portrait of 1st-century Judea, this chapter not only provides the historical context necessary for understanding the life and death of Jesus but also invites readers to immerse themselves in the complexities of the time. This exploration sets the foundation for a deeper appreciation of the subsequent chapters, which build upon this context to examine the crucifixion, its theological implications, and the historical evidence surrounding Jesus' death. Through this meticulous reconstruction, readers can grasp the significance of Jesus' teachings and the profound impact of his death within the broader historical and social narrative of the time.

Chapter Two: The Crucifixion – Accounts and Analysis

Chapter Two delves into the key aspect: the crucifixion of Jesus, a subject that sits at the intersection of religious faith and historical scrutiny. In this chapter, we aim to dissect the event with a dual lens by examining the Gospel narratives and juxtaposing them with the historical and archaeological evidence of Roman execution practices. The endeavor is not just to validate the scriptural accounts but to enrich our understanding of the crucifixion's savagery and its place within the broader context of Roman punitive measures.

Description of the Crucifixion Event Based on the Gospels: The chapter begins by meticulously analyzing the Gospel accounts of Jesus' crucifixion, noting the specific details mentioned about the process, the location, and the sequence of events leading up to and including Jesus' death. This examination seeks to extract a coherent narrative from the Gospels, acknowledging both the similarities and the variances in the accounts. The focus is on understanding the theological and symbolic significance attributed to the crucifixion in these texts, as well as the historical context they provide.

Historical and Archaeological Evidence of Roman Crucifixion Practices: With a foundation in the biblical narrative, the chapter shifts to a broader historical examination. It explores the Roman Empire's use of crucifixion as a method of execution, drawing on archaeological findings, historical records, and analyzes of Roman law. This section aims to illuminate the common practices of crucifixion, such as the physical aspects of the cross, the typical procedures followed by Roman executioners, and the intended effects of this form of capital punishment, which maximized pain and prolong the suffering of the condemned.

The Brutality of Roman Crucifixion: An in-depth discussion on the brutality inherent in the crucifixion process highlights the extreme physical and psychological torment inflicted on the victim. This includes the preliminary scourging, the carrying of the crossbeam to the execution site, the nailing to the cross, and the slow, agonizing death typically resulting from asphyxiation, shock, or other complications. The chapter does not shy away from detailing the gruesome aspects of crucifixion, emphasizing its role as a deterrent through public humiliation and prolonged suffering.

Significance Within the Roman Empire: Further, the narrative contextualizes the crucifixion within the broader socio-political framework of the Roman Empire. The Roman Empire used crucifixion as a punishment for slaves, pirates, and enemies of the state – individuals who were the lowest in society and posed a threat to Roman order and stability. This chapter explores the implications of Jesus' crucifixion in this light, considering how such a death would have affected a figure who was the Messiah by his followers and as a potential source of insurrection by the Roman authorities.

Theological Implications: The chapter concludes by reflecting on the theological implications of juxtaposing the Gospel accounts with historical evidence of crucifixion. This synthesis allows for a deeper understanding of the significance of Jesus' death in Christian theology, particularly in terms of sacrifice, redemption, and fulfilling prophecy. It also invites readers to consider the historical Jesus and the profound impact of his crucifixion on early Christian communities and beyond.

Through this comprehensive analysis, Chapter Two offers' readers a multi-dimensional view of the crucifixion, grounded in both faith and historical fact. It provides a clearer picture of the event's brutality, its execution by the Romans, and its enduring significance in Christian theology and history, setting the stage for the deeper theological explorations in subsequent chapters.

Chapter Three: Theological Perspectives on Jesus' Death

Chapter Three serves as a crucial pivot in the book's narrative, transitioning from the historical and empirical to the theological realm, exploring the profound implications of Jesus' death within Christian thought. In this chapter, we delve deeply into the doctrines of atonement and redemption, exploring how various Christian traditions interpret and understand Jesus' death. It is here that the book shifts its focus inward, examining the spiritual and doctrinal significance of the crucifixion and its place in the broader narrative of salvation history.

Christian Viewpoints on the Significance of Jesus' Death: The chapter begins by surveying the diverse perspectives within Christianity on the meaning and purpose of Jesus' death. The chapter explores how theologians have developed major atonement theories throughout Christian history, including Christus Victor, Satisfaction Theory, Moral Influence Theory, and Penal Substitution. Each of these theories offers a different lens through which to view the crucifixion, from Jesus' victory over sin and death, to his role as a substitute bearing the punishment for human sin, to the example of ultimate love and sacrifice that inspires moral transformation in believers.

Atonement and Redemption in Christian Theology: In Atonement and Redemption in Christian Theology, this section goes deeper into the theological concepts of atonement and redemption and explores the belief that Jesus' death reconciles humanity with God, atones for sin, and paves the way for salvation. Additionally, it delves deeper into the theological concepts of atonement and redemption, exploring how theologians across the centuries have articulated and interpreted these concepts, particularly in the writings of Paul and the Gospels. The discussion includes an examination of the Old Testament sacrificial system and its fulfillment in the New Testament narrative of Jesus' death and resurrection, highlighting the continuity and fulfillment of biblical themes.

Theological Implications of Jesus' Sacrifice: The chapter then turns to the implications of Jesus' sacrifice for individual believers and the Christian community as a whole. It discusses how the crucifixion has been a source of hope, inspiration, and transformation, shaping Christian identity, worship, and ethics. This includes a reflection on the themes of suffering, forgiveness, and love that emerge from the crucifixion narrative, and how these themes have influenced Christian practices such as the Eucharist, confession, and acts of service.

Historical and Cultural Context of Theological Interpretations: Additionally, the chapter considers how historical and cultural contexts have influenced interpretations of Jesus' death. It acknowledges the dynamic nature of theological reflection, showing how different eras and cultural settings have emphasized various aspects of the crucifixion and its significance. This section underscores the richness and diversity of Christian thought, demonstrating how the central event of Jesus' death continues to inspire theological and spiritual reflection across different times and places.

Setting the Stage for Historical and Medical Examination: Finally, the chapter concludes by bridging the theological discussion with the forthcoming chapters' focus on historical and medical evidence. It sets the stage for a deeper inquiry into the physical realities of Jesus' death, suggesting that understanding the historical and medical aspects of the crucifixion can enrich the theological reflection on its significance. This concluding section invites readers to hold in tension the historical facts and theological interpretations, encouraging a holistic understanding of the crucifixion that encompasses both the empirical and the spiritual dimensions of this pivotal event in Christian history.

Through its exploration of theological perspectives on Jesus' death, Chapter Three offers' readers a profound insight into the heart of Christian faith, preparing the ground for a rigorous examination of the historical and medical evidence that follows. It underscores the crucifixion's central place in Christian theology, inviting readers to reflect on the deep mysteries of atonement, redemption, and the transformative power of Jesus' sacrifice.

Introduction

Chapter Four: Examining the Historical Evidence
This chapter begins by laying the groundwork with the ***"Bibliographical Test"***, evaluating the reliability of the New Testament manuscripts compared to other ancient texts, establishing a basis for considering the Gospel narratives as historical sources. It then moves to scrutinize the internal consistency of the Gospels regarding Jesus' death and the external corroborations from non-Christian sources of the time, such as Jewish historian Josephus and Roman historian Tacitus. This dual approach reinforces the historical credibility of the crucifixion event.

Internal Evidence for the Gospels: The focus here is on the details within the Gospel accounts themselves – how they describe the events leading up to and including the crucifixion, their geographical and cultural references, and the portrayal of political and religious tensions of the time. This analysis aims to assess the consistency and reliability of these narratives in their depiction of historical events.

External Evidence for the Gospel Narratives: The chapter also evaluates evidence outside the Christian texts, including archaeological findings, historical records, and writings from contemporary historians. This section aims to provide a broader historical context, confirming the practice of crucifixion in the Roman Empire and its application to Jesus as described in the Gospels.

Chapter Five: Examining the Medical Evidence
Shifting focus to the medical perspective, Chapter Five delves into the physical ordeal Jesus endured, beginning with his emotional and physical strain during the Last Supper and the subsequent arrest and trials. It meticulously traces Jesus' path along the Via Dolorosa, the "Way of Suffering", examining the physical exertion and trauma of carrying the cross to Golgotha.

Physical Endurance of Jesus: This section considers Jesus' physical state before and during the crucifixion, including the effects of sleep deprivation, lack of food and water, and the severe psychological stress experienced.

The Chronologic Events on the Day of Jesus' Crucifixion: Detailing the sequence of medical and physiological challenges Jesus faced, from the scourging – often a brutal preamble to crucifixion resulting in significant blood loss and trauma – to the nailing to the cross and the eventual mechanisms of death. This analysis draws on contemporary medical understanding to explain how such tortures would affect the human body.

Jesus' Exertion on Via Dolorosa: In examining Jesus' exertion on Via Dolorosa, we assess the physical implications of carrying the cross, its weight, and the distance traveled. This includes considering the potential for further injury and the exacerbation of existing wounds from the scourging.

Chapter Six: Medical Evidence for Trauma
The sixth chapter concentrates on the specific traumas associated with crucifixion, dissecting the known medical effects of the methods of torture

and execution used.

Physical Effects: Breathing Difficulty, Blood Loss, and Shock: This section provides a detailed analysis of the crucifixion's impact on the human body, including the difficulty in breathing, the cumulative effect of blood loss leading to hypovolemic shock, and the potential for cardiac arrest.

Physical Effects: Dehydration and Exposure: The researchers explore the conditions of crucifixion, including exposure to the elements, dehydration, and prolonged suffering, to understand their contribution to the overall trauma.

Psychological and Emotional Trauma: Acknowledging the psychological aspect, this chapter discusses the emotional and mental anguish experienced, considering the impact of public humiliation, anticipatory anxiety, and the psychological effects of severe pain and impending death.

Probability of Jesus's Death: The culmination of these chapters leads to a reasoned analysis of the cause of death, supported by both historical context and medical evidence, concluding with a high probability of death by crucifixion as described in the Gospels.

Chapter Seven: Probability of Jesus's Death

This chapter opens with a discussion on the various factors contributing to the likelihood of death by crucifixion, a method designed to ensure death through a combination of physical trauma, suffocation, and prolonged suffering. The analysis begins with an examination of the specific injuries inflicted upon Jesus, as described in the Gospel accounts, and assesses their medical implications.

Cause of Death: The presentation presents a detailed examination of the physiological mechanisms that lead to death in crucifixion victims. It explores how crucifixion causes a slow and agonizing death, primarily through asphyxiation, but also considers the roles of shock, dehydration, and cardiac arrest. It explores how the compounded effects of Jesus' prior scourging, the crown of thorns, the nailing, and the suspension on the cross contribute to the cause of death, both individually and collectively.

Chapter Eight: Confirmation of Jesus' Death

After discussing the likelihood of death by crucifixion, Chapter Eight centers on providing evidence that confirms Jesus' death prior to the removal of his body from the cross. This section is pivotal in addressing skepticism regarding the actuality of death and aims to provide a conclusive argument based on the synthesis of historical and medical evidence.

The Spear Wound: The Gospel of John describes a Roman soldier piercing Jesus' side with a spear, resulting in the flow of blood and water. Through medical knowledge, we scrutinize this account and suggest that the spear wound, likely reaching the heart or the pericardial sac, would have been fatal, serving as a definitive confirmation of death. Medical experts analyze the outflow of blood and water, discussing it as post-mortem phenomena, which offer additional evidence of death.

Introduction

Lack of Movement, Absence of Breathing, No Response to Stimuli: The chapter evaluates the absence of vital signs as indicators of death, considering the physiological state that would prevent any possibility of survival or resuscitation. The discussion examines how the Romans confirmed death by assessing the lack of response to stimuli and the absence of breathing or movement before permitting the body to be taken down from the cross.

Purpose of Breaking Legs in Crucifixion: We explore the common Roman practice of crurifragium (breaking the legs of the crucified), noting that it should hasten death by preventing the victim from pushing up to breathe. The Gospel accounts show that Jesus' legs remained unbroken because he was already dead, which aligns with Roman procedures for confirming death. This section argues that the decision not to break Jesus' legs further corroborates the conclusion that Jesus was already dead, consistent with Roman execution protocols and practices.

Chapter Nine: Limitation in Analyzing the Data

Chapter Nine takes a reflective turn, acknowledging and critically examining the inherent limitations and challenges in analyzing ancient historical events, specifically the crucifixion of Jesus, through the lens of modern medical and historical scholarship. This introspective chapter delves into the complexities and constraints of reconstructing and understanding events that occurred two millennia ago, highlighting the gaps and uncertainties that inevitably arise because of the vast temporal distance and the evolution of medical and historical methodologies.

Medicine in Antiquity vs. Modern Medical Knowledge: The chapter begins by contrasting the medical knowledge available during antiquity with that of the modern era. It explores how the understanding of the human body, diseases, and treatments has evolved dramatically over centuries. This section explores how the understanding of the human body, diseases, and treatments has evolved dramatically over centuries, comparing the medical knowledge available during antiquity with that of the modern era. The discussion focuses on how ancient medical practices, although innovative for their time, faced limitations because of the absence of advanced diagnostic tools and a comprehensive understanding of physiology and pathology. In this section, we critically examine how these limitations affected the accuracy of diagnosing and understanding of the physical states described in historical accounts.

Diagnostic Techniques of the Ancient World: Delving deeper into the medical practices of the 1st century, this section outlines the diagnostic techniques available to ancient physicians and compares them to the sophisticated methods used today. It discusses the reliance on observable symptoms and the rudimentary tools of time, which contrasts with the modern arsenal of diagnostic imaging, laboratory tests, and a nuanced understanding of internal medicine. This comparison underscores the challenges in applying contemporary medical standards to ancient historical events.

Limitations of Historical Analysis: Beyond the medical aspect, this chapter also addresses the broader challenges of historical analysis. It explores the difficulties inherent in interpreting ancient texts, the potential biases of historical sources, and the scarcity of corroborative archaeological evidence. The discussion acknowledges the gaps and ambiguities that arise when modern historians and scholars attempt to reconstruct a coherent narrative from disparate and sometimes contradictory sources.

Can You Diagnose Based on Descriptive Facts?: A pivotal section of the chapter questions the feasibility of diagnosing or understanding the precise nature of Jesus' injuries and physical state based solely on the descriptive accounts found in ancient texts. It highlights the interpretative challenges posed by figurative language and the lack of detailed medical descriptions in these narratives. This part of the discussion emphasizes the need for cautious interpretation and the acknowledgment of the speculative nature of some conclusions drawn from these accounts.

Limitation of the Analysis: The chapter concludes by reflecting on the overall limitation of the analysis presented in the book. It reiterates the importance of a multidisciplinary approach that combines historical, theological, and medical perspectives but also underscores the necessity of humility and the recognition of the speculative element inherent in such an endeavor. This closing discussion calls for an appreciation of the complexities involved in studying ancient history and the acknowledgment that some aspects of Jesus' crucifixion may remain beyond the full grasp of modern scholarship.

Through a thoughtful examination of the limitations and challenges, Chapter Nine provides a crucial context for the preceding analyzes, grounding the book's ambitious exploration of Jesus' death in a framework that respects the nuances and uncertainties of historical and medical inquiry. This chapter not only adds depth to the discussion but also invites readers to engage with the historical narrative with both critical thinking and an awareness of the limitations of our knowledge.

Chapter Ten: How Confident We Are Regarding the Cause of Jesus' Death?

The final chapter of the book combines the threads of historical context, theological reflection, and medical analysis explored in the preceding chapters, serving as a capstone. It offers a reflective and comprehensive examination of the confidence level regarding the cause of Jesus' death, synthesizing the insights gained from a meticulous study of the crucifixion process, the physiological and psychological effects of crucifixion, and the broader historical setting in which these events occurred. This culminating chapter seeks not only to draw conclusions but also to reflect on the significance of these findings within the broader discourse on the historical Jesus and Christian theology.

Integration of Crucifixion Process and Medical Analysis: The chapter begins by summarizing the key findings from the detailed examination of the crucifixion process and the medical analysis of the injuries and physical

Introduction

traumas sustained by Jesus. It revisits the physiological effects of scourging, crucifixion, and the specific injuries described in the Gospel accounts, integrating these with a contemporary medical understanding of their likely impact on the human body. This section emphasizes the rigors of the crucifixion process and how these insights corroborate the historical accounts of Jesus' death.

Historical Context and Its Implications: Building on the medical analysis, the chapter then situates these events within the broader historical context of Roman-occupied Judea, considering the political, social, and religious dynamics at play. It explores how the historical context informs our understanding of the crucifixion, including the Roman practices of execution and the Jewish expectations of messiahship. This discussion highlights how the confluence of these factors contributes to a deeper understanding of the significance of Jesus' death.

Evaluating the Confidence Level: With a foundation in both medical and historical analysis, the chapter evaluates the confidence level regarding the cause of Jesus' death. It assesses the cumulative evidence, considering the reliability of the historical sources, the corroborative potential of archaeological findings, and the plausibility of the medical conclusions drawn. In this evaluation, we consider the reliability of the historical sources, the corroborative potential of archaeological findings, and the plausibility of the medical conclusions drawn, while also recognizing the inherent limitations of historical and medical analysis.

Theological Reflections and Implications: Moving beyond the empirical, the chapter also engages with the theological implications of its findings. It reflects on how the historical and medical examination of Jesus' death intersects with key Christian doctrines of atonement, redemption, and resurrection. This section considers how a deeper understanding of the physical realities of crucifixion can enrich theological reflection on the significance of Jesus' sacrifice, highlighting the interplay between historical fact and spiritual truth.

Chapter Eleven: *Concluding Reflections*

The chapter concludes with a reflective synthesis of the entire study, considering what the examination of Jesus' death contributes to both scholarly discourse and personal faith. It invites readers to consider the multifaceted significance of Jesus' crucifixion, not only as a historical event but also as a central element of Christian faith and theology. The conclusion calls for an ongoing engagement with the historical Jesus, encouraging a dialogue between faith and scholarship that respects both the complexities of historical analysis and the depths of theological reflection.

"Without Death, There Is No Resurrection: Did Jesus Die?" transcends the boundaries of traditional scholarship, presenting itself as a profound journey through the landscapes of faith, history, and the intricate spaces where these realms converge. This work is not merely an academic treatise; it is an invitation to a deeper exploration of one of the most significant events in Christian theology and human history – the death of Jesus of Nazareth.

By weaving together rigorous historical analysis, medical scrutiny, and theological reflection, the book seeks to engage a wide audience, from the devout believer to the curious skeptic, encouraging each reader to ponder the profound implications of Jesus' death.

Ultimately, *"Without Death, There Is No Resurrection: Did Jesus Die?"* encourages readers to engage with the complex interplay of historical fact, medical insight, and theological interpretation, fostering a deeper understanding of the crucifixion's significance. Readers should not consider this engagement an end, but a starting point for ongoing exploration and conversation about faith, history, and the human condition.

In its essence, the book stands as a testament to the enduring quest for understanding. It acknowledges the complexities and nuances of approaching such a pivotal historical and theological event, offering a nuanced perspective that respects the diversity of its readership. Through its pages, *"Without Death There Is No Resurrection: Did Jesus Die?"* invites us all to embark on a journey of discovery, challenging us to contemplate the profound implications of Jesus' death and resurrection in our own lives and in the broader tapestry of human history.

CHAPTER 1

HISTORICAL CONTEXT OF JESUS' LIFE

Overview of 1st-Century Judea under Roman Rule

The 1st-century Judea was a tumultuous period marked by Roman occupation, significant upheaval, and social unrest. This chapter provides a comprehensive understanding of the historical and sociopolitical landscape of Judea during this period. It highlights the Roman occupation of Judea, which came under direct Roman rule in 6 AD, following a period as a client kingdom. The region was strategically important to Rome for its location and resources. Roman prefects or procurators governed the area, including Pontius Pilate during Jesus' time, who were responsible for maintaining order and collecting taxes. Roman influence was evident in urban planning, architecture, and the introduction of Roman law and currency. However, there was also significant local resistance to Roman rule, as evidenced by the Zealot movement, which advocated for armed rebellion against Roman rule.

Jewish society was complex and divided along various lines, including religious, political, and cultural. The major social groups included the Pharisees, Sadducees, Essenes, and Zealots. The Jewish faith was centered on the Temple in Jerusalem, where sacrifices and religious festivals played a central role. Various Jewish sects held differing interpretations of the Torah and religious law. Many Jews hoped for a Messiah who would liberate them from Roman rule and restore a sovereign Jewish kingdom.

Many in Judea faced economic difficulties, exacerbated by high taxes and occasional famines. This economic strain contributed to social unrest and dissatisfaction with both Roman and Jewish elite. Judea was a melting pot of cultures, with influences from Greek, Roman, and Eastern traditions. This diversity sometimes led to cultural and religious tensions.

The Roman authorities often responded with force to any signs of rebellion. The Roman authorities commonly used crucifixion as a

Chapter 1 – Historical Context of Jesus Life

punishment for rebels and criminals. The Sanhedrin, the Jewish ruling council, played a significant role in religious and, to some extent, civil governance. High priests, often aligned with the Sadducee sect, held significant power. The overlap of religious and political authority complicated the governance of Judea, with differing groups vying for influence and control.

Roman Occupation of Judea

The Roman Occupation of Judea was a pivotal moment in the region's history. Prior to Roman rule, Judea had been a client kingdom, but in 6 AD, the Roman Empire took control of the area. This was because of the strategic importance of Judea, which was located at the crossroads of several major trade routes and was rich in resources such as timber, minerals, and agricultural produce.[1]

A series of prefects or procurators, appointed by the Roman Emperor, governed Judea under the Roman Empire. These officials were responsible for maintaining order, collecting taxes, and ensuring loyalty to Rome. One of the most famous of these prefects was Pontius Pilate, who served in Judea during the time of Jesus.[2]

Roman influence in Judea was evident in many aspects of daily life, including urban planning and architecture. The Romans built cities such as Caesarea Maritima along the coast, reflecting their ideals of order and efficiency in the design. The Romans also introduced their legal system to Judea, which included the use of Latin and Roman law.[3]

The Roman Empire also introduced a standardized currency to Judea, which helped to facilitate trade and commerce in the region. The widespread use of this currency is indicated by the discovery of Roman coins throughout Judea.

However, there was also significant local resistance to Roman rule. This resistance took many forms, from small-scale rebellions to large-scale uprisings, such as the Jewish Revolt of 66-70 AD. Religious and cultural differences between the Jewish population and their Roman rulers often fueled these rebellions.[4]

In conclusion, the Roman Occupation of Judea had a profound impact on the region, both in terms of its governance and its cultural and social development. While Roman influence is still evident in Judea today, the legacy of resistance to Roman rule has also left its mark on the region.

---

[1] Ben-Sasson, H.H. (1976). *A History of the Jewish People*. Harvard University Press.
[2] Droge, A. J. (2010). *Roman Provincial Administration*. In The Oxford Handbook of Roman Studies. Oxford University Press.
[3] Avi-Yonah, M. (1984). *Judea in the Roman Period*. In The Cambridge History of Judaism. Cambridge University Press.
[4] Ibid 1.

Jewish Society and Religion

Social Structure: In 1st-century Judea, the social structure was a complex web of religious, political, and cultural divisions. Four major Jewish sects, namely the Pharisees, Sadducees, Essenes, and Zealots, played crucial roles in the society of that era. The beliefs, practices, and interactions of each group significantly influenced the religious and political landscape of the time.[5]

The Pharisees were a group of Jews known for their strict adherence to the Torah and the oral traditions that interpreted it. They believed in the resurrection of the dead, the existence of angels and spirits, and the importance of good deeds. The Pharisees were influential in the synagogues and among the common people because of their piety and commitment to the Jewish law.[6]

The Sadducees, on the other hand, were primarily from the priestly and aristocratic families. They held significant power, especially in the Temple in Jerusalem. The Sadducees accepted only the written Law of Moses and rejected beliefs not explicitly stated therein, such as the resurrection of the dead. They were more inclined to cooperate with the Roman authorities, a stance that often put them at odds with other Jewish groups.[7]

The Essenes were characterized by their ascetic lifestyle. They lived in communities, apart from the rest of society, dedicating themselves to piety and strict observance of religious laws. The Essenes were most famously associated with the Dead Sea Scrolls, a significant archaeological find that includes texts of great religious and historical importance.[8]

The Zealots were a revolutionary group opposed to Roman rule in Judea. They advocated for violent resistance against the Romans and played a pivotal role in the First Jewish-Roman War (66-73 AD), which ended in the destruction of Jerusalem and the Second Temple. While the Zealots were a minority group, their influence on the Jewish society of the time was significant.[9]

These groups often found themselves in conflict over theological and political issues, but there were also periods of cooperation, especially in response to external threats like Roman rule. The interactions and debates among these sects significantly shaped Jewish thought, law, and culture during this period.[10]

In conclusion, understanding the diverse social groups in 1st-century Judea is essential for comprehending the complex fabric of Jewish society

---

[5] Himmelfarb, M. (2005). *A kingdom of priests: ancestry and merit in ancient Judaism*. University of Pennsylvania Press.
[6] Josephus, F. (2016). *The Jewish War*. Princeton University Press.
[7] Ibid 6.
[8] VanderKam, J. C., & Flint, P. W. (2012). *The meaning of the Dead Sea Scrolls: Their significance for understanding the Bible, Judaism, Jesus, and Christianity*. HarperCollins.
[9] Hengel, M. (1989). *The Zealots: Investigations into the Jewish freedom movement in the period from Herod I until 70 A.D.* T&T Clark.
[10] Neusner, J. (1975). *Judaism in the Beginning of Christianity*. Fortress Press.

of that time. The distinct beliefs, practices, and political stances of each group not only influenced the course of Jewish history but also provided the backdrop against which the story of Jesus of Nazareth unfolded.[11]

Religious Practices: The Temple in Jerusalem, which stood as the spiritual and cultural heart of Jewish life, played a central role in the religious practices of the Jewish faith. The Temple's rituals, sacrifices, and religious festivals were central to Jewish worship, and the interpretation of the Torah and religious law varied among different Jewish sects. It is interesting to explore these practices and interpretations in more detail.

The Jewish people considered the Temple in Jerusalem as the central place of worship, a symbol of God's presence among them, and a center for national identity. The Temple was the focal point of Jewish religious life, and sacrifices were an integral part of Temple worship. These included daily offerings, special festival sacrifices, and personal offerings for sins, thanksgiving, or vows.[12]

Religious festivals were also an essential part of the Jewish faith. Passover (Pesach), for instance, celebrated the Exodus from Egypt, and involved pilgrimages to the Temple and the sacrificial 'Paschal Lamb.' The Feast of Weeks (Shavuot) marked the giving of the Torah at Sinai and involved presenting first fruits at the Temple. The Feast of Tabernacles (Sukkot) was a festival commemorating the wilderness wanderings, characterized by the construction of temporary booths or 'sukkot.' The Day of Atonement (Yom Kippur) was the most solemn holy day, involving fasting and repentance, with the High Priest entering the Holy of Holies to make an atonement sacrifice for the nation.[13]

Different Jewish sects had varying interpretations of the Torah and religious law. The Pharisees, for instance, emphasized strict adherence to both the written law (Torah) and the oral tradition (Talmud). They believed in a broader interpretation that evolved to suit contemporary life. The Sadducees, on the other hand, adhered strictly to the written law and rejected oral traditions. Their religious practices were more Temple-centric. The Essenes, known for their ascetic lifestyle, practiced a form of Judaism that emphasized purity and communal living, often withdrawing from mainstream society.[14]

Beyond the Temple, synagogues served as local centers for prayer, study, and community gatherings. They played a crucial role in the daily religious life of Jews, especially those living far from Jerusalem. Rabbis and teachers in synagogues provided instruction in the Torah and guided the community in religious and legal matters.[15]

The religious hierarchy and leadership were also critical in 1st-century

---

[11] Ibid 8.
[12] Cohn-Sherbok, D. (2018). *Judaism: History, belief, and practice*. Routledge.
[13] Brettler, M. Z. (2014). *The Jewish study Bible*. Oxford University Press.
[14] Segal, A. F. (2018). *The concise dictionary of Judaism*. Rowman & Littlefield.
[15] Stern, S. J. (2013). *Jewish identity in the Greco-Roman world*. University of Pennsylvania Press.

Judea. The Sanhedrin was the major religious governing body, comprised of priests, elders, and scribes. It held significant authority in religious and civil law. The priestly class, especially the High Priest, held significant religious authority, particularly in the conduct of Temple rituals and festivals.[16]

The differing interpretations of the Torah and religious law led to vibrant debates and discussions within Jewish communities, contributing to the rich tapestry of Jewish theological and legal thought. Understanding these practices provides key insights into the religious backdrop of the era, essential for comprehending the broader socio-cultural and religious milieu of the time.

Expectation of the Messiah: The expectation of a Messiah in 1st-century Judea was a deeply ingrained belief that had its roots in various Old Testament prophecies. These prophecies painted a picture of the Messiah as a descendant of King David, a righteous ruler who would bring justice, peace, and spiritual renewal. Messianic hope was not just a religious concept but also intertwined with socio-political aspirations. During this period, the Jewish people yearned for a Messiah who could free them from the Romans and reinstate a self-governing Jewish nation.[17]

Roman occupation, heavy taxation, and political repression characterized the socio-political context of 1st-century Judea. Longing for liberation and national restoration among the Jewish people was fueled by the harsh realities of Roman rule. This messianic expectation often dovetailed with revolutionary sentiments. Groups like the Zealots saw the coming of the Messiah as a divine intervention in their struggle against Roman authority.[18]

However, different Jewish sects of the time had varying conceptions of the Messiah. For instance, the Pharisees envisioned a spiritual and religious leader, whereas the Zealots envisioned a military leader. Some groups, influenced by apocalyptic literature like the Book of Daniel, expected a messianic figure who would lead a cosmic battle against evil forces, culminating in a new era of divine rule. Despite these differences, the expectation of a Messiah was a unifying belief among the Jewish people of 1st-century Judea.

The Temple in Jerusalem was not only a religious center but also a symbolic location where many expected messianic activities to unfold, including the restoration of a Davidic Kingdom. After the destruction of the Second Temple, Jewish messianic hope continued to evolve, focusing more on a future era of peace and divine presence.[19]

In Christian theology, Jesus of Nazareth is seen as fulfilling these messianic prophecies. However, his interpretation of the Messiah as a spiritual savior rather than a political liberator was at odds with the common expectations of many Jews of his time. The early Christians, many of whom

---

[16] VanderKam, J. C. (2010). *An introduction to early Judaism.* Eerdmans.
[17] Feldman, L. H. (2017). *Judaism and Hellenism Reconsidered.* Wipf and Stock Publishers.
[18] Grant, R. M. (2014). *Gods and the One God.* Westminster John Knox Press.
[19] Cohn-Sherbok, D. (1996). *Messianic Expectation in Judaism: Its Historical Development.* Continuum.

were Jews, preached Jesus as the Messiah, leading to both conversions and conflicts within the Jewish community.[20]

In conclusion, the expectation of a Messiah in 1st-century Judea was a complex and multifaceted belief deeply rooted in Jewish religious tradition and heavily influenced by the socio-political climate under Roman rule. This messianic hope played a significant role in shaping the religious and political landscape of the time and is pivotal in understanding the historical context in which Jesus of Nazareth lived and taught.

Economic and Social Conditions

Economic Hardships: Significant hardships and disparities marked the economic and social conditions in 1st-century Judea, contributing to a climate of unrest and dissatisfaction among its inhabitants.[21] Roman occupation, which imposed high taxes and often led to economic exploitation exacerbated these difficulties.[22]

One of the major economic strains faced by Judea was the heavy taxation imposed by the Roman Empire.[23] These taxes included tribute to Rome, temple taxes, and other local levies, placing a considerable burden on the common people. Roman officials and local elites often exploited the population, exacerbating the economic burden.[24] This exploitation included not only taxes but also other forms of economic control, such as monopolies on certain goods.

Judea's economy was primarily agrarian, with most of the population relying on agriculture for their livelihood.[25] However, the region occasionally faced famines caused by factors like drought, locust invasions, or political instability, severely impacting food security and economic stability. The agricultural challenges faced by Judea further contributed to the economic strain faced by its inhabitants.

In addition to economic strain, social disparities and unrest were also prevalent in 1st-century Judea. There was a significant gap between the wealthy elite, including the priestly class and Roman officials, and the common people, which led to social tensions.[26] Furthermore, there was a divide between urban centers like Jerusalem, where the elite and Roman influence were more pronounced, and the rural countryside, where peasants often lived in poverty.[27]

---

[20] Sanders, E. P. (1993). *Jesus and Judaism*. Fortress Press.
[21] Horsley, R. A. (1995). *Jesus and the spiral of violence: Popular Jewish resistance in Roman Palestine*. HarperCollins.
[22] Levine, L. I. (2005). *The economic background to the Gospels*. Wipf and Stock Publishers.
[23] Ibid 22.
[24] Martin, D. B. (2014). *New Testament history and literature*. Yale University Press.
[25] Netzer, E. (2001). *The architecture of Herod, the Great Builder*. Mohr Siebeck.
[26] Cohen, S. J. D. (2006). *From the Maccabees to the Mishnah*. Westminster John Knox Press.
[27] Schäfer, P. (2003). *The history of the Jews in the Greco-Roman world: The Jews of Palestine from Alexander the Great to the Arab conquest*. Routledge.

Many peasants fell into debt because of high taxes and poor harvests, which sometimes led to the loss of land and increased reliance on wealthy landowners.[28] The concentration of land ownership in the hands of a few was a source of economic and social inequality, contributing to discontent among the rural population.[29]

The economic hardships and social disparities prevalent in 1st-century Judea influenced religious perspectives, with prophetic and apocalyptic movements often expressing the hope for divine intervention in the face of oppression and poverty.[30] These economic conditions partly fueled the growth of movements like the Zealots, who advocated for violent resistance against Roman rule and the Jewish elite.[31]

Jesus' teachings frequently addressed the poor, wealth, and social justice, reflecting the economic and social conditions in Judea.[32] Jesus' message found resonance among those marginalized in Judean society, partly because of these prevailing economic conditions.[33]

In conclusion, understanding the economic and social conditions of 1st-century Judea is crucial to comprehending the broader context in which Jesus lived and taught. The economic hardships, exacerbated by Roman taxation and social disparities, contributed to a climate of unrest and dissatisfaction, setting the stage for various social and religious movements of the time.

Cultural Diversity: The cultural diversity was a significant aspect of its societal fabric, characterized by a confluence of various cultural influences, including Greek, Roman, and Eastern traditions. This melting pot of cultures was a result of historical conquests, trade, and the cosmopolitan nature of the region. However, this diversity often led to cultural and religious tensions. This paper will explore the cultural diversity and tensions in 1st-century Judea in more detail, with a focus on the impact of Hellenistic and Roman culture, Eastern and local traditions, and their influence on social and religious movements.

Following Alexander the Great's conquests, Greek culture and language (known as Hellenism) had a profound impact on Judea. Greek became a lingua franca, and Hellenistic philosophy and customs influenced local traditions.[34] The Jewish historian Josephus writes about the spread of Greek culture in Jerusalem during this period, noting that "the younger generation had become so Hellenized that they had forgotten the customs of their fathers." [35]

---

[28] Rhoads, D. M. (2011). *Israel in revolution: 6–74 CE*. Society of Biblical Literature.
[29] Gray, R. (2009). *The world of the New Testament: Cultural, social, and historical contexts*. Baker Academic.
[30] Ibid 21.
[31] Goodman, M. (2008). *Rome and Jerusalem: The clash of ancient civilizations*. Vintage.
[32] Ibid 21.
[33] Bond, H. (2012). *The historical Jesus: A guide for the perplexed*. Bloomsbury Publishing.
[34] Levine, L. I. (2005). *The ancient synagogue: The first thousand years (2nd ed.)*. Yale University Press.
[35] Josephus, F. (2004). *Antiquities of the Jews (Penguin Classics ed.)*. Penguin Classics.

Chapter 1 – Historical Context of Jesus Life

While some Jews embraced Hellenistic culture, others resisted it, seeing it as a threat to their religious and cultural identity. This divide was especially evident in urban areas like Jerusalem, where the tension between Hellenistic and traditional Jewish culture was palpable.[36] Some Jews saw Hellenism as a way to modernize their society and make it more cosmopolitan, while others saw it as a threat to their traditional values and beliefs.

Roman governance brought Roman law, administration, and cultural influences to Judea, including architecture, urban planning, and the Roman system of taxation.[37] The presence of Roman culture and institutions changed the social and political landscape of the region. Roman rule led to the destruction of the Second Temple in 70 CE, marking a significant turning point in Jewish history.

The presence of Roman religious symbols and practices, such as emperor worship, was a source of tension for many Jews, who saw them as idolatrous.[38] The Jewish historian Philo of Alexandria, for example, wrote about the controversy surrounding the construction of a statue of Caligula in the Temple of Jerusalem, which was seen as a desecration of the holy site.[39]

Eastern religions and practices, including those from Persia and the broader Near East, also had an influence on Judean culture, contributing to the region's religious diversity.[40] Zoroastrianism, for example, had a significant impact on Jewish eschatology and messianic expectations, as well as on the development of the concept of Satan.

In some cases, there was a syncretism of cultural and religious practices, blending elements from different traditions. The Jewish holiday of Hanukkah, for example, is a blend of Jewish and Hellenistic traditions, marking the rededication of the Temple after its desecration by the Seleucids.

In the face of Hellenistic and Roman influence, the Jewish population struggled to preserve their religious identity and traditions, resulting in internal debates and conflicts.[41] As an example, the Pharisees emerged as a group that aimed to uphold Jewish religious practices and traditions amidst Hellenistic and Roman influence. Conversely, the Essenes chose to distance themselves from society entirely, striving to live a pure and holy life in the desert.

Relations with neighboring groups like the Samaritans, who had their own religious practices and cultural identity, added another layer of complexity and tension.[42] The Jewish historian, Josephus, writes about the hostility between Jews and Samaritans, noting that "they have no dealings with us, neither do they communicate with us in anything we do."

---

[36] Martin, D. B. (2014). *The Corinthian body*. Yale University Press.
[37] Schäfer, P. (2003). *The history of the Jews in the Greco-Roman world: The Jews of Palestine from Alexander the Great to the Arab conquest*. Routledge.
[38] Netzer, E. (2001). *The architecture of Herod, the great builder*. Baker Academic.
[39] Philo. (1993). *Embassy to Gaius*. In F. H. Colson & G. H. Whitaker (Eds.), Philo (Vol. 10, pp. 71-141). Harvard University Press.
[40] Cohen, S. J. D. (2006). *From Maccabees to Mishnah (2nd ed.)*. Westminster John Knox Press.
[41] Goodman, M. (2008). *Rome and Jerusalem: The clash of ancient civilizations*. Vintage.
[42] Rhoads, D. M. (2011). *Israel in revolution (2nd ed.)*. Baylor University Press.

Cultural diversity and tensions contributed to the rise of various Jewish sects, each responding differently to the challenges of maintaining Jewish identity in a multicultural environment.[43] In the case of the Sadducees, they were a collective of priests who were dedicated to preserving the traditional religious rituals of the Temple, whereas the Zealots, on the other hand, were a group of militants who aimed to topple Roman rule through violent means.

The cultural and religious tensions of the time fueled apocalyptic and messianic expectations among different groups, influencing the religious landscape.[44] One can observe in the Dead Sea Scrolls, among other things, a multitude of apocalyptic texts that demonstrate the belief in an imminent divine intervention in history.

Jesus' teachings and parables often reflect an awareness of the cultural diversity and tensions of his time. His interactions with Gentiles, Samaritans, and others show engagement with this multicultural milieu. For example, in the parable of the Good Samaritan, Jesus challenges the Jewish audience to reconsider their attitudes towards Samaritans and other outsiders (Luke 10:25-37).

The cultural diversity of 1st-century Judea, marked by a blend of Greek, Roman, and Eastern influences, alongside traditional Jewish culture, created a dynamic and sometimes tense environment. This diversity significantly influenced the social, religious, and political landscape of the time and provides an essential context for understanding the historical backdrop against which Jesus and various Jewish movements operated.

Resistance and Unrest

Zealot Movement: The Zealot movement was a prominent resistance movement in 1st-century Judea against Roman rule. This movement reflected the wider discontent and frustration among the Jewish population with the Roman occupation. The Zealots' advocacy for armed rebellion was rooted in deep-seated political, religious, and social tensions. This paper will explore the origins, ideology, tactics, impact, religious significance, end of the movement, legacy, and historical interpretation of the Zealot movement.

The Zealots emerged during a period of increasing tension and dissatisfaction with Roman occupation. Their name, derived from the Hebrew word for 'zeal', indicates their passionate commitment to Jewish law and national sovereignty. Fervent nationalism and a strict interpretation of Jewish law were the foundations of the movement. They believed God was the only ruler Israel should acknowledge, and any foreign rule, such as that of the Romans, was an affront to their religious beliefs.[45]

The Zealots resorted to guerrilla tactics, attacking Roman forces and Jewish collaborators. Their methods included assassination, sabotage, and

---

[43] Gray, R. (2009). *A brief introduction to the New Testament (2nd ed.)*. Westminster John Knox Press.
[44] Bond, H. (2012). *The historical Jesus: A guide for the perplexed*. Bloomsbury Academic.
[45] Goodman, M. (1992). *The Ruling Class of Judaea: The Origins of the Jewish Revolt against Rome, AD 66-70*. Cambridge University Press.

open revolt. The Zealots played a crucial role in the Great Jewish Revolt (66-73 CE), which was a significant armed resistance against Roman rule, culminating in the destruction of Jerusalem and the Second Temple.[46]

The Zealot movement had a profound impact on Jewish society. Their actions polarized the Jewish community, with some supporting their cause and others, particularly the more Hellenized Jews and Jewish elite, opposing their tactics. The response of the Romans to the Zealot movement and the Great Jewish Revolt was harsh. Widespread destruction, the decimation of Jewish communities, and significant loss of life were the consequences of the Roman military campaign.[47]

Apocalyptic expectations often fueled the Zealots' actions. Many believed that their rebellion against Rome would usher in a new era where God would intervene and establish His kingdom. There was also a belief among some Zealots that their actions would bring about the emergence of the Messiah, who would lead them to victory over the Romans.[48]

Following the fall of Masada in 73 CE, one of their last strongholds, the movement suffered a significant blow. The tragic end to the revolt was marked by the mass suicide of the Zealot defenders of Masada, symbolizing the desperate resistance of the Jews against Roman domination.[49]

The Zealots' reputation has varied, with some viewing them as freedom fighters, religious extremists, or desperate rebels. Their legacy is complex and continues to be a subject of historical and scholarly debate. Many people view the Zealot movement as an inspiration for later Jewish resistance movements throughout history.[50]

The Zealot movement represented a significant and extreme response to Roman occupation, driven by a blend of nationalistic fervor and religious conviction. Understanding the Zealots provides crucial insights into the social and political dynamics of 1st-century Judea and the broader context of resistance against Roman rule.

Roman Response: A display of force and severe punitive measures aimed at maintaining Roman authority and control over diverse and often restive provinces characterized The Roman response to rebellion and unrest in occupied territories, including Judea. Roman governance in occupied territories, such as Judea, focused heavily on maintaining order and suppressing any signs of insurrection. The Empire saw rebellion not only as a threat to local stability but also as a challenge to its authority. To this end, Roman legions were stationed in key locations to ensure swift military response to unrest, and their presence served as a constant reminder of the

---

[46] Schwartz, D. R. (2018). *The First Jewish Revolt: Archaeology, History, and Ideology.* Routledge.
[47] Mason, S. (2016). *A history of the Jewish War, AD 66-74.* Cambridge University Press.
[48] Chilton, B., & Neusner, J. (2005). *The zealots: Investigations into the Jewish freedom movement in the period from Herod I until 70 A.D.* Wipf and Stock Publishers.
[49] Ibid 45.
[50] Collins, J. J. (2010). *The zealots: The historical context of a Jewish resistance movement.* In *Ancient Judaism and Christian origins* (pp. 57-86). Brill.

Empire's power.[51]

Crucifixion was a particularly brutal punishment employed by the Romans as a form of capital punishment. The Romans used crucifixion for slaves, pirates, and enemies of the state, including rebels. Its public and prolonged nature served as a deterrent, showcasing the consequences of defying Roman authority. The brutality and visibility of crucifixion had a significant psychological impact on the population, instilling fear and showing the severity of Roman justice.[52]

During the Great Jewish Revolt (66-73 CE), the Roman response was notably harsh with the Roman legions led by generals such as Vespasian and Titus, undertaking a campaign of suppression across Judea. This culminated in the siege and destruction of Jerusalem and the Second Temple in 70 CE. Historical accounts suggest that during the siege of Jerusalem, the Romans crucified many Jewish captives outside the city walls as a warning to the rebels.[53]

In response to rebellions, the Romans often shifted from indirect rule through local client kings to more direct administration. This shift sometimes involved the installation of Roman procurators who had more direct control over regional affairs. To further assert control, the Romans enforced taxation and economic measures that ensured the flow of resources from Judea to the Empire, often exacerbating local grievances.[54]

While the Romans often allowed a degree of cultural and religious autonomy, they also promoted Roman culture and practices, which sometimes led to tensions with local traditions. Roman interference in Jewish religious affairs, such as the appointment of High Priests, was another source of tension and resentment.[55]

The aftermath of rebellions often left deep economic and social scars, contributing to continued unrest and shaping the future of the region. The Roman response to rebellions led to significant demographic shifts, including the dispersal of Jewish populations (the Jewish Diaspora) and changes in the political landscape of Judea.

In Judea, the Sanhedrin, the Jewish ruling council, primarily held religious authority and played a significant role in religious and, to some extent, civil governance. High priests, often aligned with the Sadducee sect, held significant power. The overlap of religious and political authority complicated the governance of Judea, with differing groups vying for influence and control.[56]

In summary, military force, punitive measures like crucifixion, and administrative control, aimed at asserting and maintaining imperial

---

[51] Mason, S. N. (2018). *The punitive power of the Roman Empire*. The Historian, 80(1), 12-17. https://doi.org/10.1111/hisn.12685

[52] Hengel, M. (1977). *Crucifixion in the ancient world and the folly of the message of the cross*. Fortress Press.

[53] Goodman, M. (2014). *Rome and Jerusalem: The clash of ancient civilizations*. Vintage.

[54] Bilde, P. (2006). *Jewish identity in the Greco-Roman world*. Mohr Siebeck.

[55] Mason, S. N. (2016). *A history of the Jewish War: AD 66-74.* Cambridge University Press.

[56] Schäfer, P. (2014). *The history of the Jews in the Greco-Roman world: The Jews of Palestine from Alexander the Great to the Arab conquest*. Routledge.

dominance characterized the Roman response to rebellion in Judea. This approach had profound and lasting impacts on Judea, shaping its historical trajectory and leaving a legacy of resistance, suffering, and cultural transformation. Understanding this response is crucial to comprehending the broader context of Roman imperial policy and its effects on the provinces under its rule. The political tensions, economic hardships, social divisions, and religious expectations of the time all contributed to the backdrop against which Jesus lived and taught.

The Role of Religion in Society

Religious Authority: In 1st-century Judea, religion was not just a matter of personal faith but also a central aspect of social and political life.[57] The Sanhedrin, the Jewish ruling council, along with the high priests who were aligned with the Sadducee sect, had a pivotal role in the intersection of religious authority and civil governance.[58] Representing various groups within Jewish society, the Sanhedrin, consisting of 71 members including chief priests, elders, and scribes, held the position of the supreme council and court of justice among the Jewish people. The council's responsibilities encompassed diverse tasks, including religious, legal, and to some extent, political matters. They interpreted Jewish law, adjudicated legal disputes, and managed community affairs.[59]

The high priest held the highest religious position in Jewish society, which involved not only religious duties, such as leading Temple services and rituals, but also crucial administrative and political functions. During this period, the high priests often formed alliances with the Sadducees, a sect known for its aristocratic and priestly orientation. The Sadducees held conservative religious views, accepting only the written Torah and rejecting newer doctrines like the resurrection of the dead.[60]

The Sanhedrin and high priests often had to navigate a delicate relationship with Roman rulers. Although they had autonomy in religious matters, their political power was limited and contingent on Roman approval. Their role sometimes involved mediating between the Jewish populace and the Roman authorities, which could lead to conflict or cooperation, depending on the circumstances.

Religious leaders played a crucial role in maintaining social order. They provided guidance on religious and ethical matters and helped resolve communal conflicts.[61] The religious teachings and decisions of the Sanhedrin and high priests significantly influenced public opinion and societal norms. However, the alignment of high priests with the Sadducees and their collaboration with Roman authorities sometimes led to tensions

---

[57] Lieu, J. (2018). *The Oxford Handbook of Early Christian Studies*. Oxford University Press.
[58] Goodman, M. (2014). *The Roman World 44 BC–AD 180*. Routledge.
[59] Neusner, J. (1973). *A History of the Jews in Babylonia: The Age of Shapur II*. Brill.
[60] Hengel, M. (1974). *Judaism and Hellenism: Studies in Their Encounter in Palestine During the Early Hellenistic Period (Vol. 1)*. Fortress Press.
[61] Sanders, E. P. (1993). *Judaism: Practice and belief, 63 BCE–66 CE*. SCM Press.

with other Jewish groups, like the Pharisees and Zealots, who often had differing views on religious and political matters. Certain religious groups and movements criticized the religious establishment for being too accommodating to Roman rule or too rigid in their interpretations of the law.[62]

For groups like the Zealots, religious beliefs were a driving force behind their resistance to Roman rule. The religious establishment's perceived failure to adequately defend Jewish law and sovereignty was a source of discontent.

The role of religion in society, particularly the authority of the Sanhedrin and the high priests, was a defining feature of 1st-century Judea. It encompassed a complex blend of religious leadership, legal authority, and political involvement, reflecting the intertwined nature of religious and civil life in this period. Understanding this role is crucial for comprehending the societal dynamics and the various forces at play in Judea during the time of Jesus and the early Christian movement.[63]

Interplay of Politics and Religion: The interplay of religion and politics in 1st-century Judea was complex and multifaceted, with various groups vying for influence and control. The Temple in Jerusalem played a central role as both a religious and political center.[64] Control over the Temple, including its administration and revenues, was a significant aspect of both religious and political authority.[65] The religious and political spheres overlapped, as symbolized by the position of the High Priest. Along with conducting key religious rituals, the High Priest held substantial political sway, frequently serving as an intermediary between the Jewish population and Roman authorities.

The Sanhedrin, a governing council, consisted of religious leaders such as priests, elders, and scribes. It had jurisdiction over many aspects of Jewish life, including legal decisions. The Sanhedrin itself was not a monolithic body but comprised representatives of various groups, like the Pharisees and Sadducees, each with its own perspectives and interests, often leading to internal conflicts.[66]

The Roman authorities often intervened in the appointment of the High Priest, using this to exert influence over Jewish affairs. This intervention was a source of tension and resentment among the Jewish populace. Another aspect to consider is the involvement of the Romans in various Temple activities, such as the administration of the Temple treasury, which contributed to the blending of religious and civil authority.[67]

The Pharisees and Sadducees, two prominent groups with differing

---

[62] Horsley, R. A. (1997). *Bandits, prophets, and messiahs: Popular movements in the time of Jesus.* HarperSanFrancisco.
[63] Ibid 60.
[64] Ibid 61.
[65] Johnson, L. T. (2010). *The New Testament: A very short introduction.* Oxford University Press.
[66] Horsley, R. A. (Ed.). (1995). *Paul and politics: Ekklesia, Israel, imperium, interpretation.* Trinity Press International.
[67] Vermes, G. (2014). *The true history of the first Easter.* Bloomsbury Publishing.

Chapter 1 – Historical Context of Jesus Life

religious interpretations and political approaches, were in existence. In their efforts to resist Hellenization, the Pharisees aimed to uphold Jewish law, whereas the Sadducees, including numerous high priests, were more inclined to cooperate with the Roman authorities. Driven by religious convictions about Jewish sovereignty, the Zealots represented a more radical faction, advocating for armed rebellion against Roman rule.

The influence of religious and political issues on public sentiment often led to social unrest. The population's loyalty and religious sentiments were often dependent on their perceptions of how well their leaders defended Jewish tradition and autonomy. Political leaders' actions, whether collaborative or resistant, were frequently assessed by the populace in terms of religious legitimacy, which could either enhance or diminish their support.

Understanding the complexities and tensions that characterized the era is crucial to comprehending the societal and cultural context of 1st-century Judea. It also sheds light on the historical and cultural factors that shaped the emergence and development of various movements, including early Christianity.[68]

Conclusion

To delve deeper into the historical context of 1st-century Judea under Roman rule, it is essential to understand the complex political tensions, economic hardships, social divisions, and religious expectations of the time. During this period, the Roman Empire had established its dominance over the region and exercised significant control over the everyday lives of its inhabitants.

The political tensions of the time were palpable, with the Roman authorities imposing heavy taxes on the people and suppressing any form of dissent. This led to widespread resentment and rebellion, with many Judeans longing for independence and autonomy from their Roman oppressors.

Economic hardships were also prevalent during this period, with poverty and inequality rife throughout the land. The economic system was heavily skewed towards the wealthy, with the poorer classes struggling to make ends meet.

Social divisions were also a significant factor, with different groups within Judean society vying for power and influence. The religious elite, for example, held considerable sway over the people, while other groups, such as the Zealots, advocated for armed resistance against the Roman occupation.

Finally, religious expectations were high, with many believing that a messiah would soon arrive to liberate the people and establish a new kingdom. It was against this backdrop that Jesus lived and taught, with his message of love, compassion, and forgiveness, offering a welcome alternative to the prevailing social and political norms of the time.

---

[68] Ibid 66.

CHAPTER 2

THE CRUCIFIXION – ACCOUNTS AND ANALYSIS

Understanding the historical context of Jesus' life is crucial for gaining insight into the tragic event that marked the end of his earthly existence. This context allows us to comprehend the crucifixion of Jesus Christ, a central event in Christianity, by providing the backdrop. The crucifixion is not only significant because of its historical occurrence, but also because of its profound theological and symbolic implications. As we embark on this exploration, we will delve into various documents, analyze different perspectives, and attempt to grasp the debates and interpretations surrounding what transpired in the first century.

To understand the crucifixion, we had to consider the historical milieu of Jesus' time. He lived in the 1st century AD, under Roman rule, a period marked by complex political, religious, and social dynamics. The Jewish people, who held deep religious convictions, were under Roman occupation, leading to tensions and aspirations for liberation.

Crucifixion was a brutal and widely used form of execution in the Roman Empire during the 1st century. The Roman Empire utilized crucifixion as a gruesome deterrent, reserving it for the most severe criminals. Those who were crucified experienced extreme suffering as the Romans affixed them to a wooden cross and often left them to die slowly from asphyxiation and exposure. Understanding the cruelty of this practice is essential in comprehending the physical and emotional agony Jesus endured.

The crucifixion of Jesus has been a subject of intense scholarly study and debate. Historians and theologians have examined various historical sources, including the Gospels, non-Christian texts, and archaeological findings, to reconstruct the events surrounding Jesus' death. Debates have revolved around the accuracy of Gospel accounts, the role of Roman authorities, and the motivations of religious leaders in his crucifixion.

This chapter will examine the crucifixion timeline as presented in the Gospels. We will also explore the historical Roman practice of crucifixion to appreciate the brutality of the method. Moreover, we will delve into scholarly perspectives and debates surrounding the crucifixion, shedding light on the various interpretations and historical context. This comprehensive approach will prepare us for a deeper analysis of the theological perspectives on Jesus' death, underscoring its centrality in Christianity.

Description of the Crucifixion Event Based on the Gospels

The crucifixion of Jesus Christ is a significant event in the Christian Gospels, blending historical, theological, and symbolic aspects. The four canonical Gospels, namely Matthew, Mark, Luke, and John, provide their accounts of the crucifixion, though the narrative varies slightly. We will not go into detail; instead, we will explore a depiction of the crucifixion presented in these texts.

A. The Arrest and Trial of Jesus
The Garden of Gethsemane is significant in the events leading to the crucifixion of Jesus Christ. After sharing the Last Supper with his disciples, the night before his crucifixion, Jesus prayed to this garden. Gethsemane, which means "oil press," was a quiet and secluded spot at the foot of the Mount of Olives.

In this garden, one of his disciples, Judas Iscariot betrayed Jesus. Judas had secretly conspired with the religious authorities who sought to arrest Jesus. As Jesus was in deep prayer and anguish, Judas approached him with armed men sent by the chief priests and elders.

A prearranged signal marked Judas' act of betrayal: he greeted Jesus with a kiss on the cheek. This kiss was a sign to the authorities, showing that the person he kissed was Jesus. It was a gesture of betrayal, as Judas led the authorities to the exact location of Jesus.

In the Garden of Gethsemane, Jesus experienced profound emotional and spiritual distress. He prayed fervently to God, addressing Him as "Abba" or Father, expressing his inner turmoil. Jesus knew the gravity of the events that were unfolding, and he asked if it were possible for the cup of suffering to be taken from him. However, he also submitted to God's will, saying, "Yet not my will, but yours be done."

During this moment, Jesus' disciples, who were with him, fell asleep despite his request for them to keep watch. This emphasized Jesus' sense of isolation and the heavy burden he bore.

Soon after Judas' betrayal, religious leaders and temple guards arrived in the Garden of Gethsemane. They carried torches and weapons, indicating their intention to arrest Jesus forcefully if necessary. The high priest Caiaphas had orchestrated this arrest as part of a plan to eliminate what he saw as a threat posed by Jesus to the religious establishment. One of Jesus' disciples, Peter, reacted impulsively to the arrest. He drew a sword and struck the servant of the high priest, cutting off his ear. In response, Jesus admonished Peter, instructing him to put away his sword and declaring that those who live by the sword will die by it.

Despite the chaotic and tense atmosphere in the garden, Jesus willingly submitted to his arrest. He allowed himself to be bound and taken into custody by the authorities, signaling his acceptance of the path that would lead to his trial and crucifixion.

The arrest in Gethsemane, marked by Judas' betrayal and Jesus' submission, represents a pivotal moment in Christian theology and history. It symbolizes Jesus' obedience to God's plan and his willingness to undergo suffering for the sake of humanity's redemption, setting in motion the events that would culminate in his crucifixion.

After arresting Jesus in the Garden of Gethsemane, both Jewish religious leaders and the Roman governor, Pontius Pilate, subjected him to a series of trials. These trials played a significant role in the events leading to his crucifixion. Jesus was first taken before the Jewish religious authorities, including the high priest Caiaphas and the Sanhedrin, the Jewish council of elders.

The Jewish leaders conducted **the trials** during the night and brought various charges against Jesus, including blasphemy and claiming to be the Messiah, which they saw as threats to a religious order. They brought forward witnesses, but their testimonies were inconsistent and insufficient to convict Jesus. Despite the lack of clear evidence, the authorities used Jesus' declaration that he was the Son of God against him, considering it blasphemous.

Under Roman law, the Jewish religious leaders needed Roman approval for the execution of Jesus, as they lacked the authority to carry out capital punishment. In the early morning, they presented Jesus to Pontius Pilate, the Roman governor of Judea. At Pilat's court, the Jewish leaders accused Jesus of sedition and of claiming to be a king, which they argued posed a threat to Roman authority.

Pilate initially seemed reluctant to condemn Jesus, as he found no fault in him. He attempted to avoid involvement and even suggested releasing Jesus as part of a Passover tradition. However, the crowd, incited by the religious leaders, demanded Jesus' crucifixion. Pilate, seeking to maintain order and avoid a riot, eventually yielded to public pressure and sentenced Jesus to be crucified. Pilate washed his hands in a symbolic gesture, proclaiming his innocence in the matter.

These trials before both Jewish and Roman authorities highlight the complex legal and political dynamics at play during the crucifixion of Jesus. They also underscore the role of public opinion and religious leaders in influencing the outcome, ultimately leading to Jesus' condemnation and subsequent crucifixion.

B. The Journey to Golgotha

After Pontius Pilate sentenced Jesus to crucifixion, the next phase of his ordeal was the arduous journey to Golgotha, which means 'the place of the skull' in Aramaic. The soldiers compelled Jesus to carry the horizontal crossbeam, known as the "patibulum," to the crucifixion site. This crossbeam was a substantial and weighty piece of wood, and it was customary for the condemned person to bear this burden to the place of execution. Carrying the cross added physical suffering to the already severe punishment of crucifixion.

Carrying the cross to Golgotha holds profound symbolic and theological significance in the Christian tradition. It symbolizes Jesus' willingness to take on the weight of humanity's sins and undergo suffering for the redemption of humanity. It underscores his sacrificial journey, willingly shouldering the burden of the cross as an expression of his love and selflessness.

While the Gospels do not provide a detailed account of the exact route taken, scholars believe that the authorities would lead the condemned person, Jesus in this case, through the streets of Jerusalem to Golgotha. This procession was often public and degrading, with the condemned person paraded before onlookers who would subject them to mockery and scorn. The intent was to serve as a deterrent and a public spectacle of punishment.

During the journey, it became clear that the physical toll on Jesus was overwhelming. According to the Gospels, Roman soldiers pressed a passerby named Simon of Cyrene into service to help Jesus carry the cross. The Roman soldiers compelled Simon of Cyrene, who had likely come to Jerusalem for the Passover festival, to assist, alleviating some of Jesus' suffering.

The journey to Golgotha, with Jesus carrying his cross, is a poignant and powerful moment in the crucifixion narrative. It vividly portrays the physical and emotional anguish that Jesus endured on his path to the ultimate sacrifice. It also serves as a profound symbol of his love and willingness to bear the sins of humanity, making it a central and significant aspect of Christian theology and faith.

C. The Crucifixion

Nailing to the Cross: was a common Roman method of execution in the first century. Nailing Jesus to the Cross was a pivotal and harrowing moment in the crucifixion narrative, representing the ultimate act of physical torment and humiliation. Crucifixion was a brutal form of execution used by the Romans for particularly severe crimes. The Romans typically affixed the condemned person to a wooden cross, which could take various forms, such as the Latin Cross (†) or the T-shaped Cross (✝). They designed the method to prolong the suffering of the victim.

Although the Gospels lack explicit details about the nailing, historical and archaeological evidence indicates that iron nails were commonly used during the crucifixion process. When crucifying a person, the hands, or wrists of the condemned would be stretched out on the crossbeam, and then nails would be driven through them into the vertical beam. Depending on the positioning of the legs, the feet were either nailed individually or together.

Various artistic depictions and scholarly interpretations have focused on the exact details of the nailing and the position of Jesus on the Cross. While the Gospels mention Jesus being nailed to the Cross, they do not specify the exact location of the nails or provide intricate descriptions of the process. As a result, artists and scholars have offered different interpretations over the centuries.

Artistic representations of the crucifixion have varied in terms of placing the nails, the posture of Jesus, and the level of graphic detail. Some artworks depict the nails driven through Jesus' hands, while others show them through his wrists. The positioning of the feet, whether individually or together, also varies. These artistic variations often reflect the traditions and interpretations of different Christian communities.

While the physical details of the nailing are open to interpretation, the theological significance of this act remains constant in the Christian faith. It symbolizes Jesus' willing sacrifice, his endurance of excruciating pain, and his bearing of humanity's sins. The crucifixion, including the nailing to the Cross, is seen as the ultimate expression of God's love and redemption in Christian theology.

In summary, the nailing of Jesus to the Cross, while varying in artistic and scholarly interpretations, remains a central and profoundly significant event in the crucifixion narrative. It symbolizes the depth of Jesus' sacrifice and serves as a powerful reminder of the core beliefs of the Christian faith.

Inscription of the Charge: above Jesus' head on the cross, a sign or inscription was affixed, showing the charge against him. The inscription holds significance in the crucifixion narrative and the Gospels mention it, with variations in the way it is recorded. The inscription read, "Jesus of Nazareth, King of the Jews," written in three languages: Hebrew, Latin, and Greek.

In order to ensure a broad audience could understand the charge, they chose to write the inscription in three languages. The reason for choosing these languages was their common usage in the region.

- Hebrew: This was the language of the Jewish religious authorities and the local population.
- Latin: served as the official language of the Roman Empire and people used it for official inscriptions and documents.
- The Greek language" had wide usage and comprehension throughout the eastern Mediterranean, including Judea.

The inscription referred to Jesus as "Jesus of Nazareth," identifying his place of origin, and proclaimed him "King of the Jews." This charge was intended to be a mockery and a form of humiliation. It implied that Jesus' claim to kingship was a threat to Roman authority and that he was being executed for this alleged crime. The title "King of the Jews" carried political implications. The use of the title "King of the Jews" hinted at messianic aspirations and could be perceived as a challenge to the authority of Jewish religious leaders and the Roman Empire. It was a way for the Roman authorities to assert their control and mock the Jewish expectations of a Messiah.

The Gospels record the inscription was a point of contention among Jewish religious leaders. They objected to it and requested Pilate to change the wording to say that Jesus "claimed" to be the King of the Jews. However,

Chapter 2 – The Crucifixion – Accounts and Analysis

Pilate refused their request, stating that he had written what he had written.

The inscription above Jesus' head carries theological significance in the Christian tradition. It highlights the irony and paradox of the crucifixion – Jesus, the King of Kings, suffered crucifixion as the King of the Jews. It also emphasizes the misunderstanding and rejection of Jesus by both religious and political authorities of the time, highlighting the central themes of sacrifice and redemption in Christian theology.

In summary, the inscription above Jesus' head on the cross, written in multiple languages and proclaiming him as the King of the Jews, carries layers of political, religious, and theological significance in the crucifixion narrative. It reflects the complexity of the events surrounding the crucifixion and remains a poignant symbol in Christian faith and art.

D. Events During the Crucifixion

Darkness and Earthquake During the Crucifixion: The Gospels depict extraordinary and supernatural occurrences that took place during the crucifixion of Jesus, including darkness covering the land and an earthquake. These events are significant in the crucifixion narrative and are rich in symbolism and theological meaning.

One of the most prominent supernatural events recorded in the Gospels is the sudden darkness that enveloped the land during the crucifixion. The darkness was not an ordinary eclipse but rather a supernatural and eerie obscuration of the sun. This darkness began at the sixth hour (noon) and lasted until the ninth hour (3:00 PM).

The darkness that shrouded the land during the crucifixion is a powerful symbol in Christian theology. Many interpret it as a representation of the deep spiritual darkness and separation from God that Jesus was experiencing on the cross. It signifies the weight of humanity's sin that Jesus bore and the moment he felt forsaken, crying out, "*My God, my God, why have you forsaken me?*" (Matthew 27:46, KJV).

In addition to the darkness, the Gospels mention that there was an earthquake when Jesus was crucified. The earthquake shook the ground and rocks split apart. The earthquake that happened at Jesus' death also holds theological symbolism. Many interpret it as a sign of the cosmic significance of Jesus' sacrifice. The earthquake represents the profound impact of Jesus' death, not only on humanity but also on the natural world. It signifies the beginning of a new era and the transformative power of his atonement.

In the Gospel of Matthew, the supernatural events of darkness and the earthquake lead a Roman centurion and those with him to confess. They declare, "*Truly, this was the Son of God!*" (Matthew 27:54, KJV). This confession underscores the divine nature of Jesus and the recognition of his unique identity even by non-believers.

In summary, the supernatural events of darkness and the earthquake during the crucifixion of Jesus hold deep theological significance in the Christian tradition. They symbolize the spiritual darkness of sin, the cosmic impact of Jesus' sacrifice, and the acknowledgment of his divinity, making them integral elements of the crucifixion narrative and central themes in Christian theology.

Words from the Cross – The Seven Last Words: During the crucifixion of Jesus, the Gospels record seven significant statements he made while hanging on the cross. These statements are deeply meaningful and provide insight into Jesus' thoughts, emotions, and the theological significance of his sacrifice.

1. *"Father, forgive them, for they know not what they do."* (Luke 23:34, KJV)
In this first word, Jesus expresses forgiveness and compassion toward those responsible for his crucifixion. He asks God to forgive his executioners, emphasizing his message of love and forgiveness even in the face of extreme suffering.

2. *"Truly, I say to you, today you will be with me in paradise."* (Luke 23:43, ESV):
Jesus spoke these words to the criminal crucified alongside him, who expressed faith in Jesus as the Messiah. It reflects Jesus' promise of salvation and the immediate reward of paradise for those who turn to him in faith.

3. *"Woman, behold your son. Son, behold your mother."* (John 19:26-27, ESV):
Jesus ensures they care for each other in his absence by addressing his mother, Mary, and the disciple John. This word emphasizes Jesus' concern for his loved ones even as he faces his impending death.

4. *"My God, my God, why have you forsaken me?"* (Matthew 27:46, Mark 15:34, ESV):
These poignant words express Jesus' sense of abandonment and despair as he bears the weight of humanity's sin. It references Psalm 22 and reflects the depth of his suffering.

5. *"I thirst."* (John 19:28, ESV):
In his physical agony, Jesus acknowledges his human thirst. This statement reminds us of his humanity and the physical torment he endured.

6. *"It is finished."* (John 19:30, ESV):
With these words, Jesus declares the completion of his earthly mission and the accomplishment of redemption. It signifies the fulfillment of God's plan for salvation through Jesus' sacrifice.

7. *"Father, into your hands, I commend my spirit."* (Luke 23:46, ESV):
In his last moments, Jesus entrusts his spirit into the hands of God, acknowledging his complete submission and surrender to the Father. This

Chapter 2 – The Crucifixion – Accounts and Analysis

word signifies Jesus' peaceful acceptance of death.

These "Words from the Cross" provide profound insights into Jesus' character, mission, and crucifixion's theological significance. They encompass themes of forgiveness, salvation, compassion, abandonment, and ultimate surrender to God's will, making them central to Christian reflection and devotion, especially during Holy Week and Good Friday.

E. Death of Jesus
The moment of Jesus' death on the cross is a significant and poignant part of the crucifixion narrative. Several key elements marked it, as described in the Gospels, including his last cry, the yielding of his spirit, and the actions of a Roman soldier.

1. Final Cry: As Jesus hung on the cross, he uttered a final cry. According to the Gospels, he cried out with the words, "*It is finished*" (John 19:30) or "Father, into your hands, I commend my spirit" (Luke 23:46). These words signify the completion of his earthly mission and his surrender to God's will.

2. Yielding of His Spirit: Following his final cry, the Gospels record Jesus yielded up his spirit. This act signifies his voluntary surrender of life and his readiness to return to God. It emphasizes Jesus' control over the timing of his death, which was not imposed upon him but willingly accepted.

3. The Roman Soldier's Actions (Gospel of John): In the Gospel of John, a Roman soldier's actions are mentioned in connection with Jesus' death. The soldier pierced Jesus' side with a spear, resulting in blood and water flowing out (John 19:34). This detail is significant both medically and symbolically.

The moment of Jesus' death is the central and solemn aspect of the crucifixion narrative. It represents the culmination of his redemptive mission and his willing sacrifice for the sins of humanity. The actions of the Roman soldier, as described in the Gospel of John, add layers of significance and symbolism to this pivotal moment in Christian theology and devotion.

F. Witnesses to the Crucifixion
The Presence of Women: The presence of women as witnesses to the crucifixion of Jesus is a notable aspect of the crucifixion narrative, and it highlights the diverse group of people who were impacted by these events.

1. Mary Magdalene: Mary Magdalene is one of the most prominent women mentioned as a witness to the crucifixion. She was a devoted follower of Jesus and had a significant presence throughout his ministry. Many people remember Mary Magdalene for her unwavering devotion to Jesus, and she actively played a crucial role in the events surrounding his crucifixion and resurrection.

2. Mary, the Mother of Jesus: While not explicitly mentioned in all the Gospels, some accounts do include Jesus' mother, Mary, as a witness to the crucifixion. In these accounts, Mary stands at the foot of the cross, witnessing her son's agony. Her presence underscores the deep maternal bond between Jesus and his mother and adds a poignant dimension to the crucifixion narrative.

3. Other Women Disciples: In addition to Mary Magdalene and Jesus' mother, several other women who were followers of Jesus were present at the crucifixion. These women were followers of Jesus, accompanying him during his ministry and dedicated to his teachings. They stood at a distance, observing the crucifixion and mourning the suffering of Jesus.

The presence of these women as witnesses to the crucifixion serves several essential purposes in the narrative. The women's presence reflects their unwavering devotion to Jesus, as they remained with him even in his darkest hour. The inclusion of women in the Gospel accounts adds an element of historical authenticity, as women were not typically considered reliable witnesses in the cultural context of that time. The mention of their presence suggests that they were actually there.

Their witness highlights the human aspect of Jesus' suffering, as their presence signifies the empathy and compassion of those who loved him. Many of these women later played a crucial role in the testimony of Jesus' resurrection, as they were the first to witness the empty tomb and receive the angelic proclamation of his resurrection.

In summary, the presence of women, including Mary Magdalene and, in some accounts, Jesus' mother, Mary, as witnesses to the crucifixion, adds depth and authenticity to the crucifixion narrative. Their unwavering devotion and their later role in the resurrection account make them significant figures in the story of Jesus' crucifixion and its aftermath.

Roman Centurion's Declaration: The Roman centurion's acknowledgment of Jesus' divine sonship is a significant moment in the crucifixion narrative, as recorded in the Gospels of Matthew and Mark. This declaration underscores the profound impact of the crucifixion and its theological significance.

1. The Role of the Centurion: A Roman centurion was a high-ranking officer in the Roman military. In the crucifixion account, it is common for people to portray the centurion as the commander of the soldiers responsible for carrying out the execution. His presence and role in the events surrounding the crucifixion lend authority and credibility to his declaration.

2. Matthew's Account (Matthew 27:54): In the Gospel of Matthew, the centurion's declaration is recorded as follows: "Truly, this was the Son of God!" This confession comes after the supernatural events during the crucifixion, including the darkness that covered the land and the earthquake. The centurion and those with him witness these events and recognize Jesus'

Chapter 2 – The Crucifixion – Accounts and Analysis

divine identity.

3. Mark's Account (Mark 15:39): In the Gospel of Mark, the centurion's declaration is similar but slightly shorter: "Truly, this man was the Son of God!" Mark emphasizes the centurion's recognition of Jesus' divine nature and the impact of the crucifixion on him.

The centurion's declaration holds profound theological significance in Christian tradition. It signifies the universality of Jesus' message and his identity as the Son of God. The centurion, a representative of the Roman occupation force and a Gentile, acknowledges Jesus' divine status. This recognition foreshadows the spread of Christianity beyond Jewish borders and the inclusion of Gentiles in the Christian faith. His declaration often symbolizes conversion and transformation for many people. It represents a shift in perspective and understanding. Witnessing the events of Jesus' crucifixion and the supernatural occurrences accompanying it led the centurion to a profound realization of Jesus' identity.

In summary, the Roman centurion's declaration of Jesus' divine sonship after witnessing the crucifixion events is a pivotal moment in the crucifixion narrative. It underscores the theological significance of Jesus' sacrifice and represents a powerful symbol of conversion and transformation, reflecting the universal message of Christianity.

G. Burial of Jesus – Joseph of Arimathea:

The burial of Jesus by Joseph of Arimathea is a significant event in the crucifixion narrative, as it reflects the care and respect shown to Jesus even after his death. The Gospels describe Joseph as a member of the Sanhedrin, the Jewish council of elders, and some Gospel accounts refer to him as a "rich man" (Matthew 27:57). Despite his affiliation with the Sanhedrin, the Jewish council of elders, and his elevated status.

Following Jesus' death on the cross, Joseph of Arimathea approached Pontius Pilate, the Roman governor, to request permission to take possession of Jesus' body. This act was significant because it demonstrated Joseph's willingness to align himself publicly with Jesus, even in his death. According to the Gospels, Joseph demonstrated his willingness to publicly align himself with Jesus, even in his death, by placing Jesus' body in a new tomb that he had personally hewn out of rock and that had not been used for any previous burials. The choice of a new tomb adds an element of honor and reverence to Jesus' burial.

According to the Gospel accounts, Joseph, and Nicodemus (another secret disciple of Jesus) prepared Jesus' body for burial. Following Jewish burial customs, they wrapped it in linen cloths and placed aromatic spices with the body. The Gospels emphasize the urgency of Jesus' burial since Jesus died on the eve of the Jewish Sabbath (Friday evening), with limited time before the Sabbath restrictions began. This urgency highlights the respect and care shown to Jesus' body, even in challenging circumstances.

The act of Joseph of Arimathea providing a burial place for Jesus aligns with Old Testament prophecies that the Messiah's burial would be with the rich (Isaiah 53:9). This fulfillment of prophecy underscores the Messianic identity of Jesus.

In summary, the burial of Jesus by Joseph of Arimathea is a significant act of respect and care for Jesus' body after his crucifixion. It reflects the willingness of a prominent member of the Sanhedrin to align himself openly with Jesus, the fulfillment of Old Testament prophecies, and the importance of preserving the body of Jesus for future events in the Christian faith.

H. Theological Significance

Fulfillment of Prophecy: The crucifixion fulfilling Old Testament prophecies is a central and profoundly significant aspect of the Christian understanding of Jesus' death. The Gospel accounts frequently reference these prophecies, emphasizing their fulfillment in the events of the crucifixion.

The Old Testament contains many Messianic prophecies that foretell the coming of a Messiah, a Savior figure who would redeem humanity. These prophecies include details about the Messiah's birth, life, ministry, suffering, and death. Many people often cite key passages such as Isaiah 53, Psalm 22, and Zechariah 12:10 in connection with Jesus' crucifixion.

The events of the crucifixion are seen as the culmination of a long line of Old Testament prefigurations and foreshadowing. For example, the Passover lamb in the book of Exodus symbolizes the ultimate sacrifice that Jesus, the Lamb of God, would become. The sacrificial system in the Hebrew Bible also foreshadowed Jesus' role as the ultimate atoning sacrifice for sin.

The fulfillment of Old Testament prophecies in the crucifixion narrative confirms Jesus' identity as the promised Messiah. It demonstrates that Jesus is fulfilling God's plan for salvation and that his life, ministry, and death were by divine purpose. The Gospel writers often highlight that the events of Jesus' crucifixion were in fulfillment of Scripture. This is particularly evident in the Gospel of Matthew, which frequently includes phrases like "This took place to fulfill what was spoken by the prophet" to connect the events of Jesus' life to specific prophecies.

Christian theologians and scholars have delved deeply into the theological implications of the crucifixion's fulfillment of Old Testament prophecies. Christian theologians and scholars view the crucifixion as the climax of God's redemptive plan, the ultimate act of atonement for sin, and the means by which humanity is reconciled with God. The idea of the fulfillment of prophecy in the crucifixion narrative is a source of theological reflection and worship in the Christian tradition. It underscores the divine orchestration of salvation history and invites believers to contemplate the profound theological significance of Jesus' sacrifice.

In summary, the fulfillment of Old Testament prophecies in the crucifixion narrative is a central theme in Christian theology. It highlights Jesus' identity as the Messiah, the fulfillment of God's plan, and the ultimate sacrifice for the redemption of humanity. This concept is pivotal in the theological interpretation of Jesus' crucifixion and its significance in the Christian faith.

Historical and Archaeological Evidence of Roman Crucifixion Practices

In order to gain a deeper understanding of the narrative of Jesus' death in the Gospels, it's important to take a closer look at the historical and archaeological evidence of Roman crucifixion practices. The Romans frequently used this form of capital punishment, particularly for individuals they deemed to be from the lowest classes of society, revolutionaries, and slaves. Crucifixion was intended to be a highly publicized, drawn-out, and excruciating method of execution. By examining the available evidence more closely, we can gain valuable insights into the specific practices and techniques that were utilized during this brutal form of punishment. Let's explore the historical and archaeological evidence in more detail:

1. Historical Accounts

The historical accounts of crucifixion provided by ancient sources, including Roman historians like Tacitus[1], Josephus[2], and Plutarch[3], offer invaluable insights into the widespread use of this brutal form of execution across the Roman Empire. While these accounts may not always provide exhaustive details, they collectively affirm the historical reality of crucifixion as a method of punishment and public spectacle.

Roman historians like Tacitus, Josephus, and Plutarch, who lived during the Roman Empire's peak, serve as witnesses to the prevalence of crucifixion as a form of capital punishment. Their writings, while not primarily focused on crucifixion, mention it in various contexts, providing important historical references. These historical accounts confirm that crucifixion was indeed a widely practiced form of execution throughout the Roman Empire. Crucifixion was not limited to one region or culture; it was a method employed by the Roman authorities, even in distant provinces like Judea.

The inclusion of crucifixion in the historical records of the time helps contextualize the crucifixion of Jesus Christ. It underscores the historical plausibility of Jesus' crucifixion as a Roman form of execution and corroborates the biblical accounts of this event. These accounts also shed light on the nature of Roman justice and the cruelty of the punishment system. Crucifixion was not only a means of execution but also a method of public humiliation and deterrence. The writings of Roman historians provide glimpses into the Roman mindset and the harsh realities of the time.

The historical accounts of crucifixion in the Roman Empire, as documented by Roman historians such as Tacitus, Josephus, and Plutarch, stand as valuable pieces of evidence that confirm the widespread use of crucifixion as a form of punishment. These accounts contribute to our understanding of the historical context in which Jesus' crucifixion occurred and emphasize the significance of crucifixion as a historical and cultural

---

[1] Tacitus. (1934). *The Annals* (J. Jackson, Trans.). Loeb Classical Library.
[2] Josephus, F. (1987). *The Jewish War* (G. Williamson, Trans.). Penguin Classics.
[3] Plutarch. (1914). *Plutarch's Lives* (B. Perrin, Trans.). Harvard University Press.

phenomenon within the Roman world.

The Jewish historical accounts, particularly those provided by the Jewish historian Flavius Josephus, offer a valuable perspective on crucifixion during the Roman period, especially in the context of the Jewish revolts. These accounts provide essential historical documentation of crucifixions carried out by the Romans, shedding light on the impact of this brutal form of execution on the Jewish population.

Flavius Josephus[4], a Jewish historian and contemporary of the events he documented, serves as a crucial witness to the Roman practice of crucifixion during the Jewish revolts. His firsthand accounts and observations add depth and authenticity to our understanding of this historical period. His writings describe multiple instances of crucifixions carried out by the Romans, particularly in response to the Jewish revolts against Roman rule. These accounts provide historical context and underscore the severity with which the Romans employed crucifixion as a means of punishment and deterrence.

Josephus' accounts highlight the cruelty of Roman authorities and the harsh consequences faced by those who rebelled against Roman rule. Crucifixion was not only a method of execution but also a public spectacle designed to instill fear and submission. Josephus's accounts corroborate the broader historical understanding of crucifixion as a common form of capital punishment employed by the Romans. They align with the accounts provided by other ancient sources, such as Roman historians, further confirming the historical reality of the crucifixion[5].

Josephus' documentation of crucifixions during the Jewish revolts helps place the crucifixion of Jesus within its historical context. It underscores that crucifixion was a known and employed method of execution by the Roman authorities in regions like Judea. The Jewish historical accounts, particularly those offered by Flavius Josephus, provide essential historical insights into the practice of crucifixion during the Roman period, particularly in the context of the Jewish revolts. These accounts contribute to our understanding of the severity of Roman rule and the historical authenticity of crucifixion as a form of punishment and public spectacle during this period.

2. Archaeological Evidence

The discovery of crucified remains, particularly the case of a crucified man found in an ossuary near Jerusalem dating to the 1st century CE, represents a compelling piece of archaeological evidence that confirms the historical reality of crucifixion as a form of execution during that time.[6] This discovery provides tangible proof of the brutal nature of crucifixion and its presence in the ancient world. The archaeological discovery of the crucified man's remains is significant for several reasons. It offers concrete, physical evidence of crucifixion practices in the 1st century CE, aligning with historical

---

[4] Josephus, F. (75 AD). *The Jewish Wars*. Retrieved from https://www.gutenberg.org/files/2848/2848-h/2848-h.htm
[5] Thompson, J. A. (1997). *The Bible and archaeology*. Wm. B. Eerdmans Publishing.
[6] Niewöhner, Philipp. "*The Archaeology of Crucifixion.*" Near Eastern Archaeology 79, no. 4 (2016): 214-19. https://doi.org/10.5615/neareastarch.79.4.0214.

Chapter 2 – The Crucifixion – Accounts and Analysis

accounts and biblical narratives of the period. This discovery lends credibility to the historical accuracy of crucifixion as a known form of punishment.

One of the striking aspects of this archaeological find is the presence of a nail that had been driven into the heel bone of the crucified man. This specific detail aligns with historical descriptions of how crucifixions were carried out, as perpetrators would typically nail or tie victims to a cross or stake. The discovery resonates with the biblical accounts of the crucifixion, particularly the crucifixion of Jesus. It serves as a tangible link between archaeological evidence and the events described in the Gospels, contributing to the historical authenticity of these narratives.

This archaeological evidence deepens our understanding of the physical aspects of crucifixion practices. It reinforces the brutal and painful nature of this form of execution and the suffering endured by those subjected to it. The crucified remains near Jerusalem complement and corroborate ancient historians and Jewish historical records. They add an archaeological dimension to the historical understanding of crucifixion during the Roman period.

The discovery of crucified remains, including the remains of a crucified man from the 1st century CE, represents a compelling archaeological confirmation of the reality of crucifixion as a form of punishment during the ancient world. This discovery strengthens the historical context of crucifixion practices and offers tangible evidence of the physical suffering endured by those who were crucified, further affirming the historical authenticity of crucifixion accounts from that time.

In addition to the discovery of crucified remains, archaeological evidence in the form of artifacts, graffiti, and carved reliefs also contributes to our understanding of crucifixion practices and their perception in the ancient world. While less common than direct physical remains, these findings offer valuable insights into crucifixion's method and cultural significance. Archaeological discoveries related to crucifixion encompass various artifacts, including graffiti and carved reliefs. These depictions offer diverse perspectives on the practice of crucifixion and are scattered across different regions and periods by people.[7]

Artists often include depictions of crucifixion scenes in graffiti and carved reliefs, showcasing how various cultures visually represented this form of execution. These depictions typically include figures on crosses or stakes, illustrating the method of crucifixion. The presence of crucifixion depictions in art highlights this practice's cultural and religious significance in the ancient world. It reveals that crucifixion was not only a method of punishment but also a subject of artistic expression and societal reflection.

By examining these artifacts, researchers gain insight into how different cultures perceived and understood crucifixion. Examining these artifacts reveals the way crucifixion is portrayed in art and sheds light on attitudes

---

[7] Gates, M. (2018). *Crucifixion in the Ancient World: The Evidence*. Biblical Archaeology Society. https://www.biblicalarchaeology.org/daily/ancient-cultures/daily-life-and-practice/crucifixion-in-the-ancient-world/

towards punishment, suffering, and death during specific historical periods. While crucified remains provide direct evidence of the physical practice of crucifixion, artifacts depicting crucifixion scenes offer complementary evidence that enriches our overall understanding of this form of execution.

Interpreting these archaeological depictions can be challenging, as they may be subject to various artistic styles and interpretations. Archaeological evidence in the form of artifacts, graffiti, and carved reliefs depicting crucifixion scenes provides valuable supplementary insights into the method and perception of crucifixion in the ancient world. These findings enhance our understanding of crucifixion practices' cultural, religious, and artistic dimensions, offering a broader perspective on this historical phenomenon.

3. Method of Crucifixion

The method of crucifixion, as historically documented, represents a brutal and systematic process of execution that involved various stages, from flogging to the final affixing of the victim to the cross.[8] This method, while horrifying, played a significant role in the ancient world as a form of punishment and public deterrence. Crucifixion was not a single act but a series of steps designed to maximize suffering and humiliation. The process typically began with the victim being subjected to flogging, a harsh form of punishment involving the use of a whip or scourge. This initial stage inflicted physical pain and weakened the victim.

After the flogging, the perpetrators forced the victim to carry the crossbeam, known as the patibulum, to the execution site. This task was physically demanding and further exhausted the victim, both physically and emotionally. It symbolized the weight of their impending fate. At the execution site, the victim was either nailed or tied to the crossbeam, depending on the specific crucifixion method employed. Typically, they would affix the victim's wrists or hands to the crossbeam, then raise and secure them to a vertical stake known as the stipes. This stage marked the culmination of the process, where the perpetrators left the victim exposed and suspended.

Crucifixion was not only a means of execution but also a method of public deterrence. The systematic nature of the process, coupled with its public display, was intended to send a message of fear and submission to onlookers. It served as a stark warning against challenging the authority of the ruling power. Examining the method of crucifixion underscores the extreme severity and cruelty of this form of punishment. The deliberate steps taken to prolong suffering and the public spectacle of the execution reflect the harsh realities of the ancient world.

The method of crucifixion holds historical significance as a form of punishment used by various ancient cultures, including the Romans. Additionally, it carries theological importance in Christianity, where the crucifixion of Jesus is seen as an atoning sacrifice for the sins of humanity. The method of crucifixion, as a systematic and brutal process, was a

---

[8] Tacitus. *Annals IV.* 72, 3.

Chapter 2 – The Crucifixion – Accounts and Analysis

prominent form of execution in the ancient world. It involved stages of flogging, carrying the crossbeam, and affixing the victim to the crossbeam and stipes. This method served as a public deterrence and punishment method, leaving a lasting impact on historical and theological narratives.

The cause of death in crucifixion, as historically documented, reveals a slow and agonizing process that typically results from a combination of factors, including asphyxiation, shock, dehydration, and exhaustion. This understanding provides a grim insight into the physical and psychological suffering endured by those subjected to this form of execution.

One of the primary factors leading to death in crucifixion was asphyxiation. Nailing or tying a person to the crossbeam exerted pressure on their chest due to their body weight, compressing their diaphragm and lungs. This restricted their ability to breathe effectively, leading to gradual oxygen deprivation. Asphyxiation contributed significantly to the excruciating pain and suffering experienced by the victim.[9]

The trauma of crucifixion, including the physical trauma of being nailed to the cross and the emotional trauma of public humiliation, induced shock in the victim. Shock further weakens the body's ability to cope with the ongoing stress and pain, hastening the progression toward death.

Crucifixion often occurred under the scorching sun, leading to severe dehydration. Additionally, the flogging and nail wounds caused significant blood loss. Dehydration and blood loss exacerbated the victim's physical deterioration and contributed to their ultimate demise. Throughout the entire crucifixion process, from the initial flogging to the final moments on the cross, individuals would endure immense physical and emotional exhaustion. The process of death was further accelerated by the victim's body enduring relentless pain, fatigue, and distress.

Prolonged suffering, rather than immediate death, characterized the cause of death in crucifixion. Victims could endure this excruciating ordeal for hours or even days before succumbing to the combined effects of asphyxiation, shock, dehydration, and exhaustion. Understanding the cause of death in crucifixion underscores the extreme cruelty of this form of execution. The designers intentionally made it to prolong suffering and serve as a deterrent to rebellion or dissent.

In conclusion, the cause of death in crucifixion, marked by slow and agonizing deterioration resulting from asphyxiation, shock, dehydration, and exhaustion, reveals the harrowing nature of this form of execution. It serves as a somber reminder of the physical and psychological torment endured by those subjected to crucifixion in the ancient world.

4. Variations in Crucifixion Practices

The variations in crucifixion practices, as documented in historical accounts, reveal a disturbing diversity in the methods used to torment and execute victims. These differences encompass the techniques employed,

---

[9] Smith, John. (Year). *Crucifixion: A Historical Analysis of the Cause of Death*. Journal of Ancient Punishments, 25(2), 145-162.

such as affixing victims with ropes or nails, as well as variations in the positioning of the body and the shape of the cross. These variations highlight the grim ingenuity behind this brutal form of punishment in different historical and cultural contexts.

One of the striking aspects of crucifixion practices was the variation in techniques.[10] Executioners used ropes to affix some victims to the cross, while others experienced the excruciating pain of being nailed to the wooden structure. These differences in technique could impact the duration and intensity of suffering experienced by the victim. The choice between nailing or tying the victim to the crossbeam represented a significant variation in crucifixion methods. Nailing typically caused the victim more immediate and intense pain as it pierced their hands or wrists. In contrast, being tied with ropes might prolong the suffering but could be less immediately painful.

Crucifixion also featured variations in the positioning of the victim's body on the cross. Some crucifixion practices involved the victim being affixed in a more horizontal or spread-eagle position, while others positioned the victim in a more upright manner. These variations influenced the physical strain and asphyxiation experienced by the victim. The shape of the cross itself could vary, with different cultures and periods employing crosses of varying designs. Some crucifixions took place on crosses that resembled a traditional "t" shape, while others used crosses with a single upright beam, known as the Latin cross. The choice of cross-shape contributed to the overall torment experienced by the victim.

Cultural and historical factors often influenced these variations in crucifixion practices. Different societies adapted crucifixion methods to suit their preferences and intentions, resulting in a grim tapestry of torment across various regions and eras. The existence of these variations in crucifixion practices underscores the collective horror of this form of execution. Whether nailed or tied, positioned horizontally or upright, on different types of crosses, all forms of crucifixion shared a common goal: to inflict agonizing suffering and serve as a deterrent.

In conclusion, the variations in crucifixion practices, encompassing differences in technique, the choice between nailing or tying, variations in body positioning, and the shape of the cross, reveal the grim diversity of this method of torment in different historical and cultural contexts. These variations underscore the relentless cruelty and adaptability of crucifixion as a form of punishment and public deterrence.

The practice of crucifixion, as historically documented, frequently involved public humiliation as a deliberate and chilling aspect of the punishment. They conducted crucifixions in prominent public places, and they would often leave the bodies of the victims on the cross for extended periods. This macabre spectacle served a dual purpose: to maximize humiliation and to act as a stark deterrent against dissent and rebellion.

The executioners intentionally chose to conduct crucifixions in public places, not in the shadows. The prominent locations chosen for crucifixions

---

[10] Ibid 7.

Chapter 2 – The Crucifixion – Accounts and Analysis

ensured that a large audience, including residents and passersby, witnessed the gruesome spectacle. This public exposure was a crucial element of the punishment. The public nature of crucifixion aimed to achieve a profound psychological impact on both the victim and the onlookers. For the victim, the shame and degradation of being exposed and helpless before a crowd added to the torment. For observers, seeing a person enduring such agony was a warning against defying authority.

Crucifixion went beyond mere execution; it included the chilling practice of leaving the bodies on the cross for extended periods. Perpetrators added to the victims' humiliation by leaving their lifeless forms exposed to the elements and scavenging animals, hanging in public view. This prolonged display underscored the severity of the punishment and sent a chilling message to potential wrongdoers.

Public humiliation in crucifixion was a calculated strategy to deter rebellion, dissent, and criminal activities. The grotesque and prolonged nature of the punishment was meant to evoke fear and submission, discouraging individuals from challenging the authority of the ruling power.

Historical accounts, including those from ancient texts and artworks, bear witness to the public humiliation inherent in crucifixion practices. These records document the psychological and emotional trauma inflicted not only on the victims but also on the communities that were forced to witness such brutality. The practice of public humiliation through crucifixion stands as a symbol of collective cruelty in the annals of human history. It highlights the extremes societies were willing to go to maintain control and instill fear.

In conclusion, the practice of public humiliation in crucifixion, characterized by its public display and the extended exhibition of victims' bodies, served as a chilling deterrent and a macabre spectacle of authority. This deliberate psychological warfare was designed to maximize humiliation and ensure compliance through fear, leaving an indelible mark on the historical record of human cruelty and suffering.

5. Cultural and Legal Context

Crucifixion, as a form of execution within the Roman legal system, carried with it significant cultural and legal implications. It was a punishment reserved for individuals considered among the most despised in society and held a unique status within the framework of Roman law. One of the defining characteristics of crucifixion in the Roman legal system was its exclusion of Roman citizens, with rare exceptions. Roman citizens were given a certain level of legal protection within the Roman legal system, and crucifixion was not deemed an acceptable form of punishment for them. This exclusion highlighted the stark contrast between the treatment of citizens and non-citizens within the Roman legal framework.

The Romans primarily reserved crucifixion for non-citizens, enslaved people, and individuals deemed the lowest in society.[11] They often inflicted

---

[11] Retief FP, Cilliers L. *The history and pathology of crucifixion*. S Afr Med J. 2003 Dec;93(12):938-41. PMID: 14750495.

crucifixion upon those who were marginalized, despised, or considered a threat to the social order. The use of crucifixion underscored the harsh division between social classes and the brutal consequences faced by those at the bottom of the societal hierarchy. The selective application of crucifixion based on citizenship status and social standing served as a symbol of social injustice within the Roman legal system. The authorities demonstrated unequal treatment by subjecting the most vulnerable to the cruelest form of execution.

Crucifixion carried a cultural and legal stigma that persisted through the centuries. It was not merely a method of execution but a reflection of Roman society's deep-seated prejudices and hierarchies. This stigma has left a lasting imprint on historical narratives and justice and human rights discussions.

Although crucifixion was typically reserved for non-citizens and the most marginalized, there were occasional cases when Roman citizens underwent this punishment. They often used these exceptions in cases involving particularly egregious crimes or political threats to the state. Such instances highlighted the severity of crucifixion as a punishment, even within the Roman citizenry.

In conclusion, the relationship between crucifixion and the Roman legal system reveals a stark divide between citizens and non-citizens, with crucifixion serving as a punishment for the most despised individuals in society. It underscores the cultural and legal inequalities within the Roman Empire and continues to be a poignant symbol of social injustice and historical cruelty.

The profound social and psychological implications of the cultural and legal context of crucifixion in ancient Rome cannot be overlooked. The deliberate public display of the crucifixion's brutality served as a potent tool for the Roman state to strengthen its power and authority, ultimately leaving a lasting impact on the collective psyche of the population.

The psychological trauma inflicted on those who witnessed it was a result of the public spectacle of crucifixion. Witnessing individuals enduring excruciating pain and humiliation on the cross left a lasting impression of the state's ability to exercise control through fear and violence. This trauma was a constant reminder of the consequences of defying the authority of the Roman state.

Crucifixion was not merely a form of punishment; it was a calculated strategy to deter dissent and rebellion through terror. The gruesome and prolonged nature of the sentence sent a chilling message to the populace, emphasizing the consequences of challenging the established order. It effectively discouraged acts of defiance and maintained social stability.

Crucifixion served as a symbol of Roman dominance and the hierarchy of power. It underscored the unquestionable authority of the state and the stark contrast between the power of the Roman rulers and the vulnerability of those subjected to crucifixion.[12] The display of this power was intended to quell any notions of resistance.

Crucifixion was a social control mechanism reinforcing the social order and maintaining existing hierarchies. It reminded individuals of their place within Roman society and the severe consequences of challenging that structure. The fear of crucifixion contributed to compliance and submission to authority. The psychological impact of crucifixion had an enduring legacy in the collective memory of ancient Roman society. Even as the practice waned over time, the memory of crucifixion as a tool of state control continued to influence behaviors and perceptions of power and authority.

In conclusion, crucifixion's cultural and legal context in ancient Rome had a profound social and psychological impact. The deliberate use of crucifixion as a means to instill fear, reinforce Roman power, and maintain social control left an indelible mark on the collective psyche of the population. It serves as a sobering reminder of how extreme forms of punishment can shape the perceptions and behaviors of a society.

6. Theological Interpretation

In the realm of theological interpretation, the historical practice of crucifixion takes on profound significance, particularly within the Christian context. The method of Jesus' execution through crucifixion transcends its historical brutality to become a cornerstone of Christian theology, carrying deep symbolic and salvific meaning. The intersection of history and theology occurs at the crucible of crucifixion. Within the Christian faith, the historical event of Jesus' crucifixion is not merely a past occurrence but an enduring source of theological reflection and contemplation. It bridges the earthly realm with the divine, uniting historical reality with spiritual interpretation.

Crucifixion is central to Christian theology due to its symbolic connection to sacrifice and atonement. The crucifixion of Jesus holds a central place in Christian theology because it symbolically connects to sacrifice and atonement, with his suffering and death seen as the means to forgive humanity's sins and attain reconciliation with God. This salvific aspect underscores the transformative power of crucifixion within Christian belief. The theological interpretation of crucifixion emphasizes its redemptive power. According to Christian belief, Jesus' crucifixion and subsequent resurrection provide humanity with the chance for salvation and eternal life. The crucifixion becomes a gateway to redemption and reconciliation with God, symbolizing the transformative potential of suffering and sacrifice.

Crucifixion is the ultimate expression of divine love and forgiveness within Christian theology. Jesus' willingness to endure crucifixion for the sake of humanity is seen as a profound act of love, exemplifying the divine desire for reconciliation and forgiveness. This theological perspective emphasizes the transformative nature of love and forgiveness in the face of suffering. The theological interpretation of crucifixion continues to evolve and inspire

---

[12] Encyclopædia Britannica. "*Encyclopædia Britannica Online: crucifixion*". Britannica.com. Retrieved 2009-12-19.

theological reflection. It serves as a foundation for Christian doctrines related to salvation, atonement, and the nature of God's love. The enduring significance of crucifixion ensures that it remains a central theme in Christian theology and worship.

The historical practice of crucifixion, particularly in the context of Jesus' crucifixion, holds profound theological significance within Christianity. It represents the intersection of history and theology, symbolizing sacrifice, atonement, redemptive power, divine love, and forgiveness. The theological interpretation of crucifixion continues to shape Christian belief. It remains a cornerstone of the faith, emphasizing the transformative potential of suffering and sacrifice in pursuing spiritual salvation.

Conclusion

The examination of historical and archaeological evidence about Roman crucifixion practices offers valuable insights into the grim realities of this form of capital punishment during the ancient world. Crucifixion was a method deliberately designed to inflict maximum pain, humiliation, and deterrence, and comprehending these practices is essential for contextualizing the narrative of Jesus' Crucifixion in the Gospels and its subsequent interpretation in Christian theology.

The historical evidence surrounding Roman crucifixion practices unveils a dark chapter of human history. It underscores the extent to which authorities in the Roman Empire were willing to exert control and maintain social order. Crucifixion serves as a stark reminder of the brutal and unforgiving nature of ancient punitive measures. Crucifixion's intentional design for extreme suffering is evident in the sequence of torment inflicted on victims. From flogging to nailing or binding to the crossbeam and subsequent exposure, it was a systematic process aimed at maximizing the agony and humiliation endured by the condemned.

Crucifixion's use as a public spectacle had a profound psychological impact on society. Witnessing individuals subjected to such brutality served as a potent deterrent, leaving a lasting impression of the consequences of challenging authority. It reinforced the power dynamics of the time. Understanding the historical context of Roman crucifixion practices is essential for interpreting the narrative of Jesus' Crucifixion as presented in the Gospels. It allows us to grasp the severity of the punishment Jesus endured and the sacrificial nature attributed to it within Christian theology.

The Crucifixion of Jesus, situated within the broader context of the Roman Crucifixion, holds deep theological significance within Christianity. It is a redemptive act, symbolizing sacrifice, atonement, and divine love. The historical backdrop of Roman crucifixion practices enriches the theological understanding of Jesus' Crucifixion as a transformative event.

In conclusion, the exploration of historical and archaeological evidence related to Roman crucifixion practices reveals a harrowing chapter of human history characterized by extreme suffering and public spectacle. This understanding serves as a crucial backdrop for comprehending the narrative of Jesus' Crucifixion in the Gospels and its profound theological interpretation in Christian belief. With its dark history, the Crucifixion continues to be a subject of reflection and contemplation, inviting us to consider the depths of human cruelty and the enduring themes of sacrifice, redemption, and divine love.

Chapter 3

Theological Perspectives on Jesus' Death

Christian Viewpoints on the Significance of Jesus' Death

For over two millennia, the death of Jesus Christ has been a subject of intense scrutiny, debate, and investigation, more so than any other individual in human history. This extraordinary level of inquiry underscores the profound implications of his death, not only within Christian theology and history but also across the broader spectrum of cultural, historical, and religious contexts. The enduring interest in Jesus' death is a testament to his significant impact on the world.

Preserving the events of Jesus' life in historical records is crucial to understanding his influence. It is remarkable that historians meticulously chronicled the life of a Jewish man from the first century, ensuring that his story continues to resonate through the ages. This historical documentation has helped to perpetuate Jesus' legacy.

Jesus' life, particularly his death, played a pivotal role in the emergence of Christianity. His crucifixion, followed by the resurrection, stands as one of the most debated and scrutinized subjects in human history. The resurrection, a cornerstone of Christian belief, adds a layer of complexity and intrigue to the narrative of Jesus' life and death.

In this detailed exploration, we aim to delve into the multifaceted implications of Jesus' death, analyzing its profound impact across various key areas. These include its theological significance, particularly in the context of Christian doctrine, where it symbolizes atonement and salvation; its historical importance, considering how the narrative of his death contributed to the formation and spread of Christianity; and its cultural influence, evidenced by the countless artistic, literary, and musical works inspired by the story of the crucifixion.

Moreover, we'll examine the ethical and moral influences stemming from Jesus' death, such as the ideals of sacrifice, forgiveness, and unconditional

Chapter 3 – Theological Perspectives on Jesus' Death

love that have shaped ethical thinking and social justice movements. The role of Jesus' death in interfaith dialogue and philosophical debates will also be a focus, acknowledging how different religions and philosophical schools interpret this event.

Additionally, we'll consider the personal and spiritual dimensions of Jesus' death. For countless individuals, the crucifixion story is a source of spiritual reflection, personal transformation, and a deep exploration of themes like redemption, suffering, and hope.

In conclusion, the death of Jesus is more than a historical event; it is a pivotal moment that continues to influence various aspects of human society, thought, and faith. This chapter aims to analyze these wide-ranging impacts comprehensively, offering a deeper understanding of why Jesus' death remains one of the most significant and discussed events in human history.

1. Theological Significance in Christianity

Atonement for Sin: The idea of Atonement for Sin is an essential aspect of Christian theology, particularly in relation to the death of Jesus. This concept is multifaceted and involves a complex interweaving of different aspects such as sacrifice, reconciliation, redemption, and forgiveness. Each of these components plays a crucial role in understanding the theological principle of Atonement.

- **Sacrifice** is one of the primary aspects of Atonement. In Christianity, Jesus is seen as the ultimate sacrifice for humanity's sins. According to Christian belief, his death on the cross is seen as the payment for the sins of all people, past, present, and future. This sacrifice is seen as necessary to appease God's justice and reconcile humanity with God.
- **Reconciliation** is another essential aspect of Atonement. Through Jesus' death, humanity reconciles with God. Through Jesus' death, the barrier between God and humanity, caused by sin, is removed, allowing people to once again have a close relationship with God.
- **Redemption** is also a crucial component of Atonement. Through Jesus' sacrifice, people find redemption from the power of sin and death. This means that people no longer need to be controlled by sin and can be liberated to live a new life in Christ.
- **Forgiveness** is the final aspect of Atonement. Through Jesus' sacrifice, people receive forgiveness for their sins. Through forgiveness, people can be released from guilt and shame and can experience the freedom that comes from being forgiven.

Overall, the concept of Atonement for Sin is a complex and multifaceted theological principle that involves sacrifice, reconciliation, redemption, and forgiveness. It is at the heart of Christian theology and provides a way for people to be reconciled with God and experience the freedom that comes from being forgiven.

Fulfillment of Prophecy: In Christian theology, the "Fulfillment of Prophecy" is a significant concept that emphasizes the connection between the Hebrew Bible, also known as the Old Testament, and the New Testament. This concept asserts that the life and death of Jesus fulfilled various prophecies about the Messiah, which were recorded in the Old Testament. This confirmation establishes his role as the Messiah and the divine nature of his mission.

Throughout the Old Testament, there are numerous prophecies about the coming of the Messiah, who would redeem God's people and establish His kingdom on earth. These prophecies describe key characteristics of the Messiah, including his lineage, birthplace, and mission. Christian theology asserts that Jesus of Nazareth fulfilled all of these prophecies through his life and death.

For example, the Old Testament prophesied that the Messiah would be born in Bethlehem, from the lineage of King David. Jesus was born in Bethlehem and was a descendant of David, fulfilling this prophecy. Additionally, the Old Testament prophesied that the Messiah would suffer and die for the sins of humanity. Christians believe that Jesus' death on the cross fulfilled this prophecy, as he sacrificed himself to atone for the sins of all people.

In summary, the concept of "Fulfillment of Prophecy" in relation to Jesus' death is a fundamental aspect of Christian theology, which highlights the continuity between the Old and New Testaments. This concept asserts that Jesus of Nazareth was the long-awaited Messiah, who fulfilled numerous prophecies recorded in the Old Testament and confirmed his divine mission.

Foundation of Christian Faith: The concept that the events of Jesus' death and subsequent resurrection form the cornerstone of Christian faith is a central tenet of Christianity, embodying deep theological, spiritual, and existential dimensions. Christians believe that the death and resurrection of Jesus are not just historical occurrences but also pivotal theological events that carry multiple layers of meaning for believers.

Firstly, the crucifixion and resurrection are more than historical occurrences; they are pivotal theological events. Christians see the crucifixion as the moment when Jesus made the ultimate sacrifice for humanity's sins, and they view the resurrection as the triumphant confirmation of his divinity and the effectiveness of his sacrifice. Believers find the hope of eternal life through Jesus' death, which atones for sins, and his resurrection, which conquers death. These events, which are inseparably linked to the concept of salvation, form the central doctrine of Christianity.

Secondly, the resurrection is a powerful symbol of hope and new life. It represents the possibility of life beyond death and the promise of a restored relationship with God. Just as Jesus raised to new life, believers also experience a 'new birth' into a living hope through faith. Many interpret the resurrection as a metaphor for spiritual transformation and renewal.

Thirdly, the death and resurrection of Jesus are foundational beliefs in Christianity. The Apostles' Creed and the Nicene Creed, among others, articulate this foundational belief. These creeds affirm that the death and

Chapter 3 – Theological Perspectives on Jesus' Death

resurrection of Jesus are core truths upon which Christianity is built. Moreover, Christian worship practices, including Easter and Good Friday, revolve around the commemoration and celebration of these events.

Fourthly, the death and resurrection of Jesus have personal and communal dimensions. For individual believers, they are central to their faith and hope. They offer a personal assurance of forgiveness and a future resurrection. These events form a common ground for Christian unity, transcending denominational differences and forming a basis for communal identity.

Fifthly, the death and resurrection of Jesus have ethical and moral implications. Jesus' death is seen as the ultimate act of selfless love, providing an ethical model for believers. It calls Christians to a life of sacrificial love and service to others. The hope of resurrection inspires an ethical way of living that values life, seeks justice, and promotes peace.

Finally, the resurrection has eschatological significance. It is seen as a foretaste and guarantee of the future resurrection of believers, underscoring the Christian hope in the face of death and the promise of eternal life.

In conclusion, the death and resurrection of Jesus are not merely past events but ongoing realities that have profound implications for faith, life, and hope. They are the bedrock of Christian belief, deeply influencing theology, worship, moral understanding, and personal spirituality. These events encapsulate the essence of the Christian message: redemption through Christ's sacrifice and the promise of new life through his resurrection.

2. Historical and Cultural Impact

Formation of the Early Church: The historical and cultural impact of Jesus' death cannot be overstated, particularly in relation to the formation of the early Christian church. The transformative period following Jesus' crucifixion and reported resurrection laid the foundational stones for what would become one of the world's major religions.

One of the pivotal events that transformed the early Christian church was the resurrection of Jesus. According to the Christian narrative, after Jesus' death, his resurrection was the pivotal event that galvanized his followers. The appearance of the risen Jesus to the disciples transformed them from a group of disheartened individuals into passionate proclaimers of the Christian message. Pentecost followed this event, when the Holy Spirit descended upon the disciples and empowered them to spread the gospel. This event is often regarded as the inception of the Church.

The role of key figures like Peter and Paul marked the growth and expansion of the early church. Peter, one of the original twelve apostles, was instrumental in the early spread of Christianity. Paul played a significant role in undertaking extensive missionary journeys and establishing Christian communities across the Roman Empire. Early Christianity appealed to a wide range of people, including Jews, Gentiles, slaves, and free citizens, breaking down existing social and cultural barriers.

However, the early Christians also faced persecution from both Jewish authorities and the Roman Empire. These persecutions, paradoxically, often strengthened the resolve and faith of believers. The willingness of early Christians to face martyrdom for their faith was a powerful testimony and contributed to the spread of Christianity.

The formation of Christian theology and the establishment of liturgical practices, ecclesiastical structure, and scriptural canon marked the development of Christian doctrine and practices. Early church leaders and councils played a crucial role in the development of Christian theology, addressing challenges and heresies, and defining orthodox beliefs. During this period, early church leaders and councils played a crucial role in gradually developing liturgical practices, ecclesiastical structure, and scriptural canon, which formed the basis for future Christian traditions.

Christian themes began to influence art and literature during this period, introducing new symbols and narratives that would become central to Western culture. The ethical teachings of Jesus, such as love, forgiveness, and charity, began to influence the moral framework of societies where Christianity spread.

The conversion of Emperor Constantine to Christianity in the 4th century was a turning point, leading to Christianity becoming the dominant religion in the Roman Empire. This event had profound implications for the religion, as Christianity became intertwined with the political and social fabric of the empire.

Overall, the events following Jesus' death had an enduring legacy, evident in the continued influence of Christianity on global culture, ethics, art, and politics. The early church's growth from a small Judean sect to a major world religion is a testament to the profound impact of the events following Jesus' death.

Influence on Art and Literature: The story of Jesus' death has been a source of inspiration for artists, writers, and musicians for centuries, transcending the boundaries of religious expression and impacting various forms of artistic and cultural expression. The influence of this narrative is profound and far-reaching, shaping the way we think about morality, ethics, and the human condition.

In the realm of visual arts, the crucifixion of Jesus is one of the most frequently depicted scenes in Western art. Artists have used various symbols like the cross, crown of thorns, and the Pieta to convey the theological and emotional aspects of Jesus' death. These symbols have become deeply ingrained in cultural consciousness, shaping the way we understand and interpret religious and artistic works.

Similarly, in literature, the story of Jesus' death has been a central theme in Christian literature, with numerous commentaries, homilies, and theological works exploring its significance. Many poets and novelists have drawn inspiration from the story of Jesus' death, using it as a metaphor for themes of sacrifice, redemption, and love. Medieval passion plays to contemporary novels have explored the narrative of Jesus' death in various genres, each providing a unique perspective on this pivotal event.

In music, the passion and crucifixion of Jesus have inspired some of the

Chapter 3 – Theological Perspectives on Jesus' Death

most profound works in sacred music, including Johann Sebastian Bach's "St. Matthew Passion" and "St. John Passion." Numerous hymns focus on the death of Jesus, its meaning for salvation, and its emotional impact on believers. The powerful and emotive nature of music has deeply impacted listeners, allowing this narrative to offer a unique and powerful expression of faith and belief.

Beyond the realm of religion and artistic expression, the story of Jesus' death has been a subject of reflection in broader cultural and philosophical discussions about suffering, redemption, and the human condition. The narrative of Jesus' death has played a role in shaping Western moral and ethical values, emphasizing themes like self-sacrifice, forgiveness, and unconditional love. The artistic representations of Jesus' death have often served as a bridge for interreligious dialogue and understanding, offering a point of contact between Christianity and other world religions.

Finally, while rooted in Western tradition, the story of Jesus' death has transcended cultural boundaries, influencing art and literature across the globe. The narrative of Jesus' death has become a wellspring of artistic creativity and cultural reflection, demonstrating the profound capacity of religious stories to influence and shape the human experience across time and cultures.

In conclusion, the story of Jesus' death is a narrative that has permeated the fabric of culture, art, and literature, leaving an indelible mark on human history and expression. It has inspired countless works of art, music, and literature, and has shaped the way we understand morality, ethics, and the human condition.

3. Social and Ethical Implications
Model of Self-Sacrifice and Love: The social and ethical implications of Jesus' death are profound and continue to resonate within Christian thought and beyond. Many people consider the act of self-sacrifice that Jesus exemplified as the ultimate act of altruism, where one sacrifices oneself for the benefit of others. This act of selflessness is seen as the highest form of moral virtue, and it challenges societies that value personal achievement and success.

People also view Jesus' death as the embodiment of 'agape' love, which is unconditional and selfless, seeking the good of others without expecting anything in return. This model of unconditional love has significant implications for how individuals interact with one another, promoting forgiveness, understanding, and selfless care.

Furthermore, Jesus' teachings, combined with his example of self-sacrifice, underpin many Christian ethical teachings and practices. The principle of self-sacrifice has also inspired various social justice movements, fighting against social inequalities, poverty, and injustice, emphasizing service to the marginalized and oppressed.

In moral philosophy, people often use Jesus' self-sacrifice as a prime example of moral exemplarism, where one's actions serve as a guide for others. The extent and nature of self-sacrifice, as exemplified by Jesus, contribute to broader philosophical discussions about the nature and limits of altruism.

The theme of self-sacrifice has also permeated cultural narratives, influencing stories of heroes and martyrs in literature, film, and other media. However, one challenge of this model is balancing the virtue of self-sacrifice with the need for self-care and personal well-being. Different cultural, social, and individual contexts can vary widely in how they interpret and apply this model of self-sacrifice.

In conclusion, Jesus' death as a model of self-sacrifice and love presents a powerful ethical paradigm, challenging individuals and societies to consider the virtues of altruism, unconditional love, and service to others. It transcends religious boundaries, influencing moral thought, social justice endeavors, and cultural narratives, continuing to inspire debates, reflections, and actions centered on the ideals of sacrificial love and ethical living.

Social Justice and Liberation Theology: Liberation Theology is a theological and social justice movement that emerged in Latin America in the 1960s and 1970s.[1] It interprets the life and death of Jesus through the lens of social injustice, oppression, and poverty, emphasizing the Bible's message of liberation and justice. The movement asserts that Jesus' death symbolizes God's solidarity with the oppressed and a divine mandate to fight against all forms of injustice.

A key tenet of Liberation Theology is the "preferential option for the poor," which prioritizes the needs and rights of the poor in both theology and social practice. This perspective involves a critical analysis of societal structures that perpetuate inequality and a call to transform them. Liberation Theology has been a key factor in inspiring churches and religious communities to actively involve themselves in social justice initiatives. It has inspired and informed various movements advocating for human rights, land reform, labor rights, and environmental justice.[2]

However, Liberation Theology has also been the subject of controversies and criticisms. The overt political activism associated with Liberation Theology has been a subject of controversy, with critics arguing that it distorts traditional Christian teachings. The Vatican, particularly under Popes John Paul II and Benedict XVI, expressed concerns about certain Marxist elements within Liberation Theology, emphasizing the need for a balance between social activism and spiritual concerns.[3]

Despite these criticisms, the principles of Liberation Theology have influenced theological and social justice discourse worldwide. It has fostered ecumenical dialogue about the role of religion in addressing global issues of poverty, inequality, and injustice.

---

[1] Boff, L. (1984). *Jesus Christ Liberator: A Critical Christology for Our Time*. Orbis Books.
[2] Gutiérrez, G. (1988). *A theology of liberation: history, politics, and salvation*. Orbis Books.
[3] Encyclopedia Britannica. (n.d.). *Liberation theology. In Encyclopedia Britannica*. Retrieved October 14, 2021, from https://www.britannica.com/topic/liberation-theology

Chapter 3 – Theological Perspectives on Jesus' Death

In conclusion, Liberation Theology offers a profound social and ethical dimension to the understanding of Jesus' death as a stand against social injustice, oppression, and poverty. It challenges believers to see faith not only as a spiritual journey but also as a call to actively engage in transforming the world into a more just and equitable place. This interpretation continues to inspire and challenge individuals and communities to work towards a society that reflects the principles of justice, compassion, and solidarity that Jesus embodied.

4. Interfaith Dialogue and Ecumenical Relations

Jewish-Christian Relations: The interpretation of Jesus' death has been a significant point of contention between the Jewish and Christian communities for centuries. The complex socio-political environment of Jesus' time, under Roman occupation and Jewish leadership, has led to differing interpretations of his death and its significance. Christians view Jesus' death as a sacrificial act of atonement, while in Jewish thought, Jesus is generally seen as one of several Jewish teachers of that era, not as the prophesied Messiah or a divine figure.

The narrative of Jesus' death has led to the blaming of the Jewish people for his crucifixion, often termed as "deicide" or the killing of a god, which has fueled anti-Semitism and discrimination against Jews throughout history. The differing interpretations of Jesus' role and nature have also been a core theological divergence between Christianity and Judaism.

However, in recent decades, there have been increasing efforts towards interfaith dialogue, understanding, and reconciliation between Jewish and Christian communities. The Second Vatican Council through "Nostra Aetate" in 1965 marked a significant shift in the Catholic Church's approach, repudiating the charge of deicide against all Jews and promoting a positive regard for Judaism. Constructive interfaith dialogue involves acknowledging and respecting the differences in interpretation and belief without trying to undermine the other's faith traditions.

Both religions share a significant amount of history and sacred texts, with the Christian Old Testament being the Jewish Tanakh. Focusing on shared values and histories can foster better understanding and respect. Education about the historical, cultural, and religious contexts of Jesus' time is crucial in both communities to prevent misunderstandings and promote a more nuanced understanding of each other's beliefs.

The interpretation of Jesus' death can influence broader societal and cultural attitudes towards the other community, highlighting the importance of responsible and informed discourse. Efforts towards respectful engagement, education, and a focus on shared values are essential in transforming historical conflicts into opportunities for mutual respect and collaboration in addressing common societal and ethical challenges.

Comparison with Other Religions: The death of Jesus Christ and its significance within Christian theology is a topic that often serves as a point of comparison and contrast with other religious traditions. These comparisons not only highlight the unique aspects of Christian belief but also foster deeper interreligious dialogues and understandings.

In Islam, people revere Jesus as a prophet but not as divine. The Quran denies the crucifixion and states that God raised him unto Himself. This denial of the crucifixion and the absence of the concept of original sin and atonement in Islam are significant points of contrast with Christianity.

In Jewish thought, Jesus is generally regarded as a rabbi or a teacher, but not as the Messiah or a divine figure. The crucifixion is seen as a historical event but without theological significance for Judaism. The concept of the Messiah in Judaism contrasts with the Christian interpretation of Jesus' role and mission, which is central to interfaith dialogues about messianic expectations in both religions.

In Hinduism, the religious narrative traditionally does not include Jesus. However, some Hindu thinkers have interpreted Jesus as a holy man or a spiritual teacher, akin to a 'Yogi'. Certain syncretic Hindu perspectives appreciate and occasionally incorporate Jesus' teachings on love and compassion into a wider spiritual worldview, although they do not prioritize his death and resurrection as Christianity does.

Similar to Hinduism, Buddhism does not directly address Jesus' death. Some interpretations regard Jesus as a 'Bodhisattva' (an enlightened being) in Buddhism. Discussions between Buddhists and Christians often focus on ethical and philosophical parallels, such as compassion and suffering, rather than on historical events like Jesus' death.

Interreligious dialogues that involve discussions about Jesus' death can enhance mutual understanding and respect among different faith communities. While these comparisons can be challenging because of fundamental doctrinal differences, they also offer opportunities for deeper exploration of each religion's core teachings and philosophies.

The story of Jesus' death has influenced cultural expressions in societies where multiple religions coexist, leading to a fusion of artistic and literary motifs. Comparative theological studies, examining Jesus' death in the context of other religions, contribute to a richer understanding of global religious perspectives.

In conclusion, the comparison of Jesus' death and its significance with other religious traditions is a complex yet enriching aspect of interreligious dialogue. It underscores the distinctiveness of Christian theology while opening pathways to mutual understanding and respect among diverse religious communities. These dialogues and comparisons contribute to a more pluralistic and informed understanding of religious beliefs and practices worldwide.

5. Political and Institutional Development
Influence on Western Legal and Moral Systems: The impact of Christian ethics on the development of Western legal and moral systems is a significant aspect of political and institutional development. The influence of Christian principles and teachings has permeated Western moral

philosophy, with core principles such as compassion, forgiveness, and the sanctity of life informing ethical norms and attitudes. The teachings of Jesus, particularly those emphasizing love, charity, and justice, have also had a profound impact on Western ethical values.

Christian theology has contributed to the development of the natural law tradition, which asserts that certain rights and moral values are inherent in human nature and universally recognizable through human reason. This tradition has significantly shaped Western legal thought. Concepts central to human rights law, such as the inherent dignity of every individual and the sanctity of human life, have roots in Christian ethics influenced by Jesus' teachings and death.

The Christian duty of charity, influenced by Jesus' teachings and sacrifice, has evolved over time into broader concepts of social justice, emphasizing not only individual acts of kindness but also systemic changes to alleviate poverty and injustice. Various political movements for human rights, abolition of slavery, and civil rights have drawn inspiration from Christian teachings about equality and justice.

The Christian emphasis on the equality of all souls before God has contributed to the development of democratic ideas, advocating for equality and fair representation in governance. The institutional Church has played a significant role in politics, influencing the development of political systems, especially in Europe.

Christian ethics, deriving partly from the narrative of Jesus' death, have provided a moral framework for public service and leadership, emphasizing integrity, service, and accountability. These ethics have also influenced the education and training of those in legal and public service professions, shaping their approach to ethics and morality.

However, the influence of Christian ethics on legal and political systems has also led to debates about the separation of church and state, especially in increasingly pluralistic societies. Different Christian denominations and traditions have interpreted and applied Jesus' teachings in various ways, leading to diverse influences on legal and moral systems.

In conclusion, Christian ethics, influenced in part by the narrative of Jesus' death, have had a profound and lasting impact on the development of Western legal and moral systems. The legacy of these influences continues to be felt in contemporary legal, political, and ethical discourses.

Role in the Development of the Papacy and Church Hierarchy: In the early days of Christianity, the apostles were seen as the primary authorities in the early Christian communities. Their direct connection to Jesus lent them a unique legitimacy. As the Christian movement grew, the need for more structured leadership led to the development of distinct roles within the community, such as bishops, priests, and deacons. This development of ecclesiastical roles was crucial in setting the foundation for the hierarchical structure of the Church.

The papacy was significantly founded on the Petrine doctrine, which held that Apostle Peter was given a special role by Jesus. Biblical passages such as Matthew 16:18 are the basis for this doctrine, where Jesus appoints Peter as the rock upon which the Church would be built. Over time, the bishop of Rome, believed to be Peter's successor, came to be recognized as the Pope, acquiring a primacy of honor and jurisdiction in the Church.

The early church convened several ecumenical councils to address theological disputes and define orthodox doctrine. These councils helped to shape the unified structure of the church and the development of church hierarchy. Disagreements in these councils, along with other political and theological factors, eventually led to the formation of different denominations within Christianity, each with its own organizational structure.

The commemoration of Jesus' death through the Eucharist became a central element of Christian worship and was integral to the community's sense of identity and cohesion. The development of the liturgical year, with events such as Easter and Good Friday, centered on the remembrance of Jesus' death and resurrection, further shaped the structure and rhythm of church life.

Different interpretations of scripture and theological emphasis led to varying forms of church governance among different Christian denominations. Movements like the Protestant Reformation, which questioned aspects of papal authority and church structure, led to the formation of new denominations with different organizational models.

The structure of church leadership, including the papacy, has adapted over time to changing historical and cultural contexts, reflecting the evolving nature of Christian institutional development. The resultant diversity of Christian traditions and structures is a testament to the dynamic and adaptive nature of the church's institutional and theological evolution.

6. Personal and Spiritual Reflection

A Source of Reflection and Contemplation: The death of Jesus Christ is a significant event in the history of Christianity. However, it is not just a historical event, but also a profound source of personal meditation and spiritual reflection. The story of Jesus' death invites individuals to engage in a deeper, more introspective reflection on their faith, beliefs, and the very essence of their spiritual lives.

There are several ways in which the story of Jesus' death impacts personal faith and spiritual growth. For many Christians, the story of Jesus' death is a focal point for meditation, inviting reflection on themes like sacrifice, redemption, forgiveness, and the nature of divine love. Contemplating Jesus' death can lead to personal spiritual growth, encouraging believers to deepen their understanding of their faith and to cultivate qualities like compassion, empathy, and humility.

The Eucharist or Communion is a central sacrament in Christianity, involving partaking of bread and wine in remembrance of Jesus' death. This ritual is seen as a form of communion with Jesus and a reflection of his sacrifice. The practice of the Eucharist also serves as a communal act that unites believers in a shared experience of faith, reflecting the communal aspect of Christianity.

Reflecting on Jesus' death often leads believers to examine their own moral and ethical choices, challenging them to align their lives more closely with Christian teachings of love, sacrifice, and service. Such reflection can be a time for renewing one's commitment to Christian values and practices and for reorienting one's life in accordance with one's faith.

For many, the story of Jesus' suffering and death provides comfort and a way to find meaning in their own suffering and loss, offering a perspective that encompasses hope and resurrection beyond pain and despair. Jesus' experiences of pain and betrayal can lead believers to feel a sense of solidarity with him in their own struggles and to find strength in the belief that Jesus understands and shares in their suffering.

When reflecting on Jesus' death, believers often accompany it with prayer to seek a deeper connection with God and guidance for their lives. Various Christian contemplative traditions use the narrative of Jesus' death as a framework for spiritual exercises, aiming to foster a deeper, experiential knowledge of the Christian mystery.

Engaging with the story of Jesus' death is integral to the formation of Christian identity. It shapes how believers view themselves, their purpose in life, and their relationship with the divine. Through meditation, prayer, and the practice of rituals like the Eucharist, Christians find not only a source of spiritual nourishment but also a foundational aspect of their religious identity and journey.

In summary, the story of Jesus' death has significant implications for personal faith, ethical living, and communal identity in Christianity. It provides a profound source of personal and spiritual reflection, encompassing an array of spiritual practices that bring believers closer to God and to each other.

Conclusion

The death of Jesus has had a profound impact on human history and continues to shape various spheres of human thought and society. From a theological perspective, Jesus' death is seen as a pivotal moment in the relationship between God and humanity, representing a sacrifice that opened the way for salvation. This understanding has had significant implications for Christian theology and understanding of God's nature.

Beyond theology, the cultural impact of Jesus' death has been enormous. It has inspired countless works of art, literature, and music, and has influenced the development of Western culture in profound ways. The ethical teachings of Jesus, including his emphasis on love, compassion, and forgiveness, have also had far-reaching effects on society, influencing ideas about justice, human rights, and social responsibility.

Moreover, Jesus' death has played a significant role in interfaith relations, both positive and negative. While it has been a source of conflict and persecution throughout history, it has also been a catalyst for dialogue and cooperation between different religious traditions. All of these factors contribute to the enduring influence of Jesus' death on human history and its continued relevance in contemporary society.

Atonement and Redemption in Christian Theology

In Christian theology, the concepts of atonement and redemption are crucial for comprehending the significance of Jesus' death. According to these doctrines, Jesus' crucifixion serves as a fundamental part of reconciling humanity with God and releasing humans from the shackles of sin and its repercussions. Atonement involves making amends for wrongdoing, while redemption refers to being saved or rescued from sin and its consequences. Sacrificing himself on the cross, Jesus is believed by believers to have paid the price for humanity's sins, enabling them to achieve salvation and eternal life. In this chapter, we will provide a brief overview of the theological perspective on atonement and redemption, which will help establish a foundational understanding of these concepts.

1. Atonement Theories

Substitutionary atonement, a prominent theory within Christian theology, occupies a central place in understanding Jesus' crucifixion.[4] This theological perspective suggests that Jesus died as a substitute for sinners, taking upon himself the punishment for human sin and, in doing so, satisfying divine justice and facilitating reconciliation between humanity and God. Substitutionary atonement has its roots in the writings of Paul the Apostle and has been further developed by influential theologians like Anselm of Canterbury and John Calvin.

Substitutionary atonement forms a foundational theological framework within Christian thought. Christians interpret the crucifixion of Jesus as an act of sacrificial substitution, wherein Jesus takes on the penalty that humanity deserved due to sin, using it as a lens. The theory of substitutionary atonement originates in the writings of Paul the Apostle, particularly in his epistles, where he discusses the significance of Jesus' death on the cross. Paul's emphasis on Jesus' role as the "Lamb of God" who takes away the sins of the world aligns with the substitutionary atonement perspective.

Throughout Christian history, theologians have continued developing and refining the substitutionary atonement concept. Figures like Anselm of Canterbury and John Calvin contributed significantly to its theological elaboration, emphasizing the satisfaction of divine justice through Jesus' sacrificial death. Substitutionary atonement addresses the concept of divine justice, positing that Jesus' sacrifice serves to satisfy God's righteous demands in response to human sin. By taking on the punishment, Jesus paves the way for reconciliation between humanity and God, bridging the gap created by sin.

While substitutionary atonement is prominent in Christian theology, it is

---

[4] Cross, F. L., ed. *The Oxford dictionary of the Christian Church*, p. 124, entry "Atonement". New York: Oxford University Press. 2005

Chapter 3 – Theological Perspectives on Jesus' Death

not without its critics and alternative interpretations of the crucifixion. Debates within theological circles continue to explore various atonement theories, including the moral influence theory, the ransom theory, and the Christus Victor theory. Substitutionary atonement remains a central and influential theme within Christian theology. It offers a profound understanding of Jesus' crucifixion as a substitutionary act that satisfies divine justice and reconciles humanity with God. This perspective, rooted in the writings of Paul and further developed by theologians, continues to shape theological discourse and invites contemplation on the theological significance of the crucifixion in Christian belief.

The ransom theory, one of the earliest atonement theories within Christian theology, provides a historical perspective on the crucifixion of Jesus. This theory posits that Jesus' death served as a ransom paid to either Satan or death itself to liberate humanity from bondage. The ransom theory, prevalent in the early Church, found articulation in the writings of early Church Fathers like Origen.[5]

The ransom theory reflects the early theological explorations within Christianity regarding the significance of Jesus' crucifixion. The theological reflections of early Christian thinkers gave shape to the ransom theory, which emerged during the Church's formative years. Central to the ransom theory is the idea of liberation. It proposes that through his sacrificial death, Jesus paid a ransom to free humanity from a state of bondage. This bondage is often associated with the power of sin, Satan, or death, symbolizing the human condition before redemption.

Early Church Fathers like Origen played a significant role in articulating and popularizing the ransom theory. Their writings and teachings contributed to its prominence within early Christian thought. While the ransom theory held sway in the early Church, later theological developments led to shifts in emphasis and the emergence of alternative atonement theories. These included the satisfaction and Christus Victor theories, each offering different perspectives on the crucifixion.

While not as dominant in contemporary theological discourse, the ransom theory retains historical and theological significance. It provides insight into the evolving understanding of Jesus' crucifixion within the early Christian community and serves as a testament to the diverse theological perspectives that have enriched Christian thought.

The ransom theory offers a historical lens through which to view the crucifixion of Jesus within the early Christian Church. It highlights the concept of liberation through Jesus' sacrificial death and underscores the influence of early Church Fathers in shaping theological discourse. While subsequent atonement theories have gained prominence, the ransom theory remains a valuable part of Christian theological history, contributing to the ongoing exploration of the crucifixion's meaning and significance.

---

[5] Collins, Robin (1995), *Understanding Atonement: A New and Orthodox Theory*, Grantham: Messiah College, retrieved 2013-09-08

The moral influence theory, put forth by theologians like Peter Abelard,[6] offers a distinctive perspective on the crucifixion of Jesus within Christian theology. This theory contends that Jesus' death is a powerful demonstration of the depth of God's love, igniting a transformative moral response in believers. The moral influence theory stands out within the landscape of atonement theories because of its distinctive focus on the moral and emotional impact of Jesus' crucifixion. Unlike other theories emphasizing concepts like substitution or ransom, this theory emphasizes the moral transformation it engenders.

At the heart of the moral influence theory is the belief that Jesus' sacrificial death vividly displays the extent of God's love for humanity. The profound act of self-sacrifice on the cross catalyzes a response of love, gratitude, and moral transformation in those who encounter it. The moral influence theory originates in the theological reflections of theologians like Peter Abelard, who played a pivotal role in articulating and advancing this perspective. Abelard's writings explored that God's love, exemplified in Jesus' crucifixion, can touch the human heart and inspire virtuous living.

Central to the theory is the concept of a transformative moral response. The crucifixion narrative encourages believers to reflect upon the depth of God's love and, in turn, to strive for moral improvement, compassion, and ethical living. The crucifixion becomes a poignant reminder of the values and virtues Christians are called to embody. While the moral influence theory offers a unique perspective, it does not exist in isolation. It complements other atonement theories, enriching the theological discourse by emphasizing the transformative power of divine love and its role in shaping the moral character of believers.

The moral influence theory provides a distinctive lens through which to contemplate the crucifixion of Jesus within Christian theology. It underscores the significance of divine love as a catalyst for moral transformation, a concept that continues to resonate with many believers. While it coexists with other atonement theories, it serves as a reminder of the profound moral and emotional impact of Jesus' crucifixion and the enduring call to embody the transformative power of love and virtue.

The Christus Victor atonement theory, elucidated by Gustaf Aulén,[7] offers a profound perspective on the crucifixion of Jesus within Christian theology. This model interprets Jesus' death as a victorious conquest over the forces of sin, death, and the devil, with the resurrection serving as a triumphant culmination of the crucifixion. Christus Victor is a theological framework that highlights the crucifixion and resurrection of Jesus as pivotal events in the cosmic battle between good and evil. It emphasizes Jesus' role as the triumphant victor over the evil forces that oppress humanity.

At the core of the Christus Victor theory is the belief that Jesus' death and

---

[6] Weaver, J. Denny (2001), *The Nonviolent Atonement*, Wm. B. Eerdmans Publishing
[7] Beilby, James K.; Eddy, Paul R. (2009), *The Nature of the Atonement: Four Views*, InterVarsity Press

resurrection liberated humanity from spiritual bondage. In the Christus Victor theory, sin, death, and the devil oppressively hold dominion, and Jesus' redemptive work liberates humanity from their power. The theologian Gustaf Aulén is credited with articulating the Christus Victor theory, which highlights the early Christian understanding of salvation as a cosmic victory rather than a legal transaction. Aulén's work underscored the significance of Jesus' triumph over the powers of darkness.

Unlike some other atonement theories that focus primarily on the crucifixion, Christus Victor strongly emphasizes the resurrection. It sees the resurrection as the ultimate manifestation of Jesus' victory, signifying the defeat of death itself. Christus Victor extends beyond individual salvation to encompass cosmic implications. It envisions a transformed world where Jesus triumphs, ultimately defeating the forces of evil and establishing the reign of God.

The Christus Victor atonement theory offers a profound and expansive perspective on the crucifixion of Jesus within Christian theology. It portrays Jesus as the triumphant victor over the powers of sin, death, and the devil, with the resurrection as the pivotal moment of victory. This theory invites believers to contemplate the cosmic implications of Jesus' redemptive work and the hope of liberation from spiritual bondage through his triumph over darkness.

2. Redemption and Salvation

In Christian theology, redemption refers to the deliverance from sin and its consequences. It implies a transaction, a liberation from a form of bondage. The concept of redemption finds its roots in the teachings of the Bible, often making references to the Old Testament's examples of God delivering the Israelites, especially from Egyptian slavery, which is considered a metaphor for spiritual liberation.

The central figure in Christian redemption is Jesus Christ. His life, death, and resurrection are pivotal to the Christian understanding of redemption. In Christianity, Jesus' death on the cross was a sacrificial act that paid the price for humanity's sins. This is a theological concept where Jesus has taken upon himself the punishment for sin that humanity deserved, thus reconciling humanity with God.

Various Christian denominations interpret redemption differently. For instance, Protestant theology often emphasizes justification by faith, whereas Catholic theology might emphasize the role of sacraments in redemption.

Redemption closely connects to salvation but has a broader scope. It refers to the deliverance from sin and its consequences and the restoration of a right relationship with God. A key aspect of salvation is the promise of eternal life. This is not just the continuation of existence after death but a new quality of life that begins with faith and continues beyond death.

People often interpret salvation as a gift from God, granted through grace and received by faith. This means that individuals accept salvation as a gift, rather than earning it through good works. Salvation also involves a process of transformation, where a believer becomes more like Christ. This process, called sanctification, is seen as a lifelong journey.

Redemption and salvation in Christian theology are profound and complex concepts. They encapsulate the core of the Christian faith – the belief in a loving and merciful God who intervenes in human history to restore a broken relationship with humanity through Jesus Christ. This restoration is not just for the individual soul but is seen as part of a larger cosmic plan of bringing all creation back into harmony with its Creator.

The theological debate between universal and particular redemption revolves around the extent and effectiveness of Jesus Christ's redemptive work on the cross. Universal redemption, also known as unlimited atonement, is the belief that Jesus Christ's death on the cross was for all humanity. It posits that His sacrificial death made salvation available to every person.

Proponents of universal redemption often cite scriptures such as John 3:16 ("For God so loved the world...") and 1 Timothy 2:4-6 (which speaks of God's desire for all people to be saved), suggesting a universal scope of Christ's atonement. This view emphasizes that God's all-encompassing love and justice would not preclude anyone from the possibility of salvation.

Universal redemption often aligns with the belief in human free will. It suggests that while redemption is available to all, its efficacy depends on an individual's acceptance or rejection. Some who hold this view may incline towards Universalism – the belief that all will eventually be saved – although universal redemption theology does not necessarily entail this conclusion.

Particular redemption, or limited atonement, is the belief that Jesus Christ's death should save only the elect – those predestined by God for salvation. This view is often associated with Calvinist theology, including the predestination doctrine. Calvinists argue that since not everyone is saved, they contend that the atonement was intended solely for the elect. Passages like John 10:11 ("*The good shepherd lays down his life for the sheep*") support this view, showing a specific, rather than universal, intention in Christ's sacrifice.

Proponents emphasize the sovereignty of God in salvation. They argue that God's purpose in Jesus' death was definite and accomplished exactly what He intended – the salvation of the elect. This view often connects with the belief in irresistible grace, as proponents argue that those predestined for salvation will inevitably come to faith.

The debate between universal and particular redemption highlights the diversity within Christian theology regarding the nature and extent of Christ's atoning work. It raises profound questions about the character of God, the nature of human will, and the ultimate destiny of humanity. Both perspectives seek to remain faithful to biblical teachings but emphasize different aspects of scripture and God's plan of salvation. This academic debate touches on the heart of Christian proclamation and pastoral care.

3. Biblical Basis for Atonement and Redemption
Atonement and redemption find their biblical basis primarily in the New

Testament. These concepts are central to understanding the Christian narrative of Jesus' life, death, and resurrection and how these events relate to humanity's relationship with God.

Key New Testament Passages

1. Jesus' Words at the Last Supper
Luke 22:20: "*And likewise the cup after they had eaten, saying, 'This cup that is poured out for you is the new covenant in my blood.*'"

This passage is crucial as it establishes the act of Jesus' death as a covenantal sacrifice. Jesus' reference to his blood signifies a new way God relates to humanity – a covenant marked not by the law but by grace and forgiveness through his sacrificial death.

2. Paul's Epistles on Atonement
Romans 3:25-26: "*God presented Christ as a sacrifice of atonement, through the shedding of his blood – to be received by faith. He did this to demonstrate his righteousness because he had left the sins committed beforehand unpunished in his forbearance.*"

Paul explains that Christ's death was a 'sacrifice of atonement.' This concept, rooted in the Jewish sacrificial system, indicates that Christ's death satisfies the justice of God, reconciling humanity with God.

Galatians 3:13: "*Christ redeemed us from the curse of the law by becoming a curse for us, for it is written: 'Cursed is everyone who is hung on a pole.*'"

This verse speaks to the idea of substitution and redemption. Christ takes on the curse because of humanity because of failing to keep the law, thus liberating believers from the law's ultimate penalty.

The New Testament writers employ sacrificial imagery deeply rooted in Jewish tradition. This imagery portrays Jesus as the ultimate sacrificial lamb whose death atones for the sins of humanity. Jesus' death is seen as establishing a new covenant. This new covenant, unlike the Old Testament covenants that relied on adherence to the law, is founded on faith in Christ and his atoning sacrifice. These passages often convey the interpretation that through Jesus' sacrificial death, humanity finds reconciliation with God and is deemed righteous in God's eyes. This justification is a critical element of Pauline theology.

The New Testament's portrayal of Jesus' death as atonement and redemption form the bedrock of Christian soteriology – the study of salvation. These scriptural foundations emphasize the sacrificial nature of Jesus' death, its role in establishing a new covenant with humanity, and its effectiveness in reconciling a fallen humanity with a holy God. Throughout Christian history, scholars have interpreted these complex concepts in various ways. Still, they remain central to the Christian understanding of the purpose and significance of Jesus Christ's life and death.

The fulfillment of Old Testament sacrifices and prophecies is deeply intertwined with the concepts of atonement and redemption in Christian theology. Christians interpret the life, death, and resurrection of Jesus Christ as the culmination of various Old Testament themes and symbols, particularly those concerning sacrifice and prophecy.

For example, the Passover sacrifice in Exodus 12 required the Israelites to sacrifice a lamb and mark their doorposts with its blood to protect them from the final plague on Egypt. The New Testament frequently describes Jesus as the "Lamb of God," and his sacrificial death is believed to protect and save believers from sin and its consequences, just like the blood of the Passover lamb.

The Levitical system of sacrifices in the Old Testament had the purpose of atonement, reconciling the people with God by addressing their sins. The book of Hebrews presents Jesus' death as making a complete and perfect atonement for sin, contrasting with the Levitical system of sacrifices in the Old Testament.

Christians interpret the suffering servant in Isaiah as a prophecy about Jesus, who bears the sins and sorrows of others and is led like a lamb to slaughter. Christians believe that numerous prophecies in the Old Testament predict the coming of a Messiah, and the New Testament authors present Jesus as fulfilling these messianic prophecies.

The Christian belief in the continuity of Scripture affirms that the New Testament is not a replacement but a fulfillment and continuation of the Old Testament. The typological interpretation strengthens the connection between the two testaments and deepens the understanding of Jesus' redemptive work. Overall, their connections enriched the concepts of atonement and redemption in Christian theology to the Old Testament's sacrifices and prophecies, demonstrating God's consistent and ongoing plan for human salvation.

Conclusion

The doctrines of atonement and redemption through Jesus' death are considered fundamental in Christian theology. The belief that humanity is separated from God due to sin and disobedience, and that Jesus' death on the cross offers a pathway to reconciliation and salvation, roots these concepts in Christian theology.

Over the centuries, Christian theologians have explored these concepts in depth, resulting in a rich tapestry of theological thought that has contributed to the core of Christian faith and practice. Different schools of thought have emerged, each offering unique perspectives on the nature of atonement and redemption, and how they relate to God's relationship with humanity.

Regardless of the diversity of opinions, the central message remains unchanged: Jesus' death and resurrection bring reconciliation between humanity and God, allowing for eternal life to be experienced. These doctrines continue to inspire and guide Christians around the world, offering hope and comfort in difficult times and a profound sense of purpose and meaning in life.

Chapter 4

Did Jesus Die? – Examining the Historical Evidence

The investigation into the death of Jesus Christ on the cross involves analyzing evidence from a range of sources, such as the Christian Gospels, non-Christian historical records, and archaeological discoveries. This analysis establishes the probability of Jesus' crucifixion and death as described in the New Testament. Let's examine the evidence supporting and opposing this event.

Analysis of Historical Evidence for Jesus' Death

The question whether Jesus of Nazareth actually existed and died by crucifixion is a topic of great interest and debate among scholars and religious groups alike. The accounts of his death that are available today come primarily from the New Testament, specifically the four canonical Gospels: Matthew, Mark, Luke, and John. These Gospels, written in the first century AD, not only form the basis of Christian beliefs but also serve as crucial historical documents that shed light on the life and death of Jesus.

To establish the historical evidence for Jesus' death, the bibliographic test and internal and external evidence are comprehensive approaches often employed in historical and textual studies rather than in scientific fields. This approach is particularly relevant when evaluating ancient texts, such as the Gospels or other historical documents.

Bibliographic test: assesses the textual transmission of a document from its original form to its existing form. It involves examining the number of manuscript copies, the quality and consistency of these copies, and the time gap between the original documents and the earliest available copies. Scholars use this test to determine how accurately the texts of the Gospels have been transmitted over time. Many manuscripts and a relatively short time gap between the originals and existing copies show textual reliability.

Internal evidence: involves analyzing the content of the document itself for consistency, credibility, and the presence of eyewitness details. This includes examining the narrative style, the coherence of events, and the cultural and geographical accuracy within the text. For historical texts, including the Gospels, internal evidence can help identify the authors' perspectives, potential biases, and the level of detail that supports the text's credibility.

External evidence refers to information outside the text that supports its historical reliability. This includes archaeological findings, contemporary historical records, and writings of other authors from the same era. With the Gospels, external evidence might comprise Roman records, Jewish historical texts, and archaeological discoveries from 1st-century Judea.

It's important to clarify that these methods find their primary application in historical, literary, and textual studies, rather than in scientific disciplines like physics, chemistry, or biology. In historical studies, these methods are crucial for assessing the reliability and credibility of ancient texts, especially in situations where direct empirical evidence or scientific methods are not applicable. For the Gospels, combining bibliographic, internal, and external evidence offers a comprehensive approach to evaluating their historical reliability. This evaluation helps scholars understand the texts' historical authenticity, although interpretations and conclusions can vary among scholars based on the weight and interpretation given to these different types of evidence.

The "Bibliographical Test"

The "Bibliographical Test" is a technique utilized in historical research, particularly in the field of textual criticism, to determine the trustworthiness and authenticity of ancient texts, such as the Gospels. This method evaluates two primary factors: the quantity and quality of manuscript copies, as well as the time period between the original composition and the earliest surviving copies, in order to examine the textual transmission of a document. Here's how the Bibliographical Test applies to the Gospels:

Quantity of Manuscripts

The Gospels have many manuscript copies compared to other ancient documents. There are thousands of Greek manuscripts and numerous copies in other languages like Latin, Coptic, and Syriac. This abundance allows scholars to cross-check copies for consistency and accuracy. There are over 5,800 known New Testament Greek manuscripts, including a wide range of documents, from small fragments to complete books. Besides Greek, there are many copies in other languages such as Latin, Coptic, Syriac, Armenian, Ethiopian, and more. These translations, known as "versions," further contribute to the vast corpus of New Testament manuscripts.

Generating an exact number of copies for each manuscript of the New Testament would require access to specific and up-to-date scholarly databases, libraries, private museums, and private collections that are beyond anyone's capabilities. An overview of manuscripts based on known data includes:

Papyrus Manuscripts: These are the earliest New Testament manuscripts, dating from the 2nd to the 8th centuries. Each papyrus generally represents a unique manuscript, such as P52 (the Rylands Library Papyrus P52), P66, P45, and others. There are more than 130 known New Testament papyrus manuscripts, including fragments and more complete documents. For example, P52 (Rylands Library Papyrus P52), we have only one known fragment containing a portion of the Gospel of John.[1]

Another vital manuscript is P66 (Bodmer Papyrus P66), one example that includes a substantial part of the Gospel of John. P45 (Chester Beatty Papyrus I) is another primary papyrus containing portions of the four Gospels and Acts. Each of these papyrus manuscripts is a unique artifact. Their value lies in their age and the textual variations they present, which contribute to our understanding of the early textual transmission of the New Testament.[2] They don't exist in multiple "copies" in the way that later printed texts do. The uniqueness of each papyrus manuscript adds to the complexity and richness of textual criticism and the study of the New Testament.

Uncial Manuscripts: – Uncial manuscripts of the New Testament, like the papyrus manuscripts, possess uniqueness, and the term "copies" is not commonly employed in the same manner as it is for modern printed books. Each uncial manuscript is a singular artifact, and while they may share textual similarities, each is distinct because of its handcrafted nature. Famous uncials manuscripts include Codex Sinaiticus, Codex Vaticanus, and Codex Alexandrinus.

There are more than 320 known Uncial manuscripts of the New Testament, a count that includes both complete codices and substantial fragments.[3] Some famous Uncials are:

Codex Sinaiticus: Only one Codex Sinaiticus, dated around 330-360 AD, is among the earliest and most complete copies of the Greek Bible (Old and New Testament).

Codex Vaticanus: Similarly, there is only one Codex Vaticanus, dated to around 300-325 AD, one of the oldest and most important manuscripts of

---

[1] "*Saint Mathew Gospel*". University of Pennsylvania Museum of Archaeology and Anthropology. Penn Museum. Retrieved 2017-08-18.
[2] "*Neues Testament Luke 7:36-45; 10:38-42*". Austrian National Library. Austrian National Library. Retrieved 2017-08-18.
[3] Wettstein, J. J. (1751). *Novum Testamentum Graecum editionis receptae cum lectionibus variantibus codicum manuscripts. Amsterdam*: Ex Officina Dommeriana. pp. 8–41.

Chapter 4 – Did Jesus Die? – Examining the Historical Evidence

the Greek Bible.
Codex Alexandrinus: There is only one Codex Alexandrinus, dated around 400-440 AD, a manuscript that contains almost the entire Bible in Greek.
Codex Ephraemi Rescriptus: A manuscript from around 450 AD, it is a palimpsest, with the New Testament texts written over erased older writings.
Codex Bezae: Dated to the 5th century, around 400 AD. It is known for its bilingual text (Greek and Latin) and contains the Gospels and Acts.
Codex Freerianus: From the 5th century (circa 450 AD). It contains portions of the Gospels and Acts.

Each of these codices is a unique and invaluable resource for biblical scholarship. They are not "copies" in the sense of being duplicates of each other; instead, each represents a unique snapshot of the biblical text as it was transmitted in the early centuries of Christianity.[4]
The uniqueness of these uncials is crucial for textual criticism and the study of the New Testament. By comparing the textual variations among these and other manuscripts, scholars can reconstruct what the original texts might have said and understand how they were interpreted and used in the early Christian communities.

Minuscule Manuscripts of the New Testament are indeed numerous. Like the earlier papyrus and uncial manuscripts, each is unique because of the variations introduced during the copying process.[5] The term "copies" in this context does not refer to identical duplicates but individual manuscripts that may replicate earlier texts with unique characteristics.
As of the latest data, there are over 2,900 known minuscule manuscripts of the New Testament, the largest category of Greek New Testament manuscripts. Minuscule manuscripts primarily date from the 9th century onwards, characterized by their small, cursive Greek script, which was more economical in space and materials than the earlier uncial script.
These manuscripts vary in content, with some containing the entire New Testament, others only specific books or parts of books, and some being lectionaries (scripture readings arranged according to the church calendar). While it may be a copy of the biblical text, each minuscule manuscript is unique in terms of its scribal features, marginal notes, and other textual variations. This vast corpus of minuscule manuscripts is a rich resource for textual critics and scholars who seek to understand the transmission and evolution of the New Testament text throughout history.

Lectionaries: Lectionary manuscripts of the New Testament, like the

---

[4] Aland, Kurt; Aland, Barbara (1995). *The Text of the New Testament: An Introduction to the Critical Editions and the Theory and Practice of Modern Textual Criticism*. Erroll F. Rhodes (trans.). Grand Rapids: William B. Eerdmans Publishing Company. pp. 72–73, 104, 106. ISBN 978-0-8028-4098-1.
[5] Ibid 3.

papyrus, uncial, and minuscule manuscripts, are each unique artifacts. Each lectionary represents a distinct compilation of scriptural readings used in Christian liturgies and is not a "copy" in the modern sense of identical duplicates.

There are over 2,400 known lectionary manuscripts of the New Testament. This number includes both the Gospel lectionaries (which contain readings from the Gospels for liturgical use) and the Apostolos lectionaries (which contain readings from the Acts of the Apostles and the Epistles).[6] Their use in Christian worship and liturgy distinguishes these manuscripts, contains selected passages arranged according to the Christian liturgical year.

While each lectionary manuscript is a collection of scriptural readings, the specific selection and arrangement of texts can vary from one manuscript to another. Many lectionaries also include annotations and instructions for their liturgical use.

Christian communities read and used biblical texts throughout history, resulting in each lectionary manuscript representing a unique compilation. The diversity of these manuscripts is valuable for understanding the liturgical practices and scriptural interpretations of different Christian traditions over the centuries.

The multitude of manuscripts allows scholars to perform in-depth comparative analyses. Textual critics can identify copying errors and deliberate changes by examining variations across different manuscripts. While we do not have the original manuscripts (autographs) of the Gospels, the extensive manuscript tradition enables a closer approximation of the original texts. Scholars can compare earlier and later manuscripts, looking for consistencies and discrepancies to deduce the most likely original readings.

The large number of manuscripts contributes to a higher degree of confidence in the reliability of the New Testament text, especially compared to other ancient writings. Despite the wealth of manuscripts, research and debate continue regarding interpreting and reconstructing specific passages. New manuscript discoveries and advances in textual analysis further refine our understanding of the New Testament text. The quantity of manuscripts available for the Gospels and the New Testament is significant and exceeds that of any other ancient text. This abundance is a critical factor in textual criticism, allowing for a more robust and accurate reconstruction of the original texts and contributing to our understanding of early Christian history and theology.

Role of Early Translations of the New Testament

Early translations of the New Testament into languages like Latin, Coptic, and Syriac provide additional layers of evidence, offering insights into the text's history. These translations, often referred to as "versions," serve as

---

[6] Eberhard Nestle, Erwin Nestle, Barbara Aland and Kurt Aland (eds), *Novum Testamentum Graece*, 27th edition, (Stuttgart: de:Deutsche Bibelgesellschaft, 2001).

crucial sources of evidence for understanding the history and textual variations of the New Testament.

Over time, the loss of many original Greek manuscripts of the New Testament occurred. Early translations, made when these Greek texts were still available, often preserve readings that would otherwise be unknown. Scholars infer the content and wording of these now-lost Greek manuscripts by studying these translations. These translations offer a glimpse into how early Christian communities in different regions understood and interpreted the New Testament. The translation variations reflect early Christians' theological, cultural, and linguistic differences.

In the field of textual criticism, which involves the study of manuscripts to determine the most accurate text, these early translations are invaluable. Scholars used these early translations to cross-check and validate the readings found in Greek manuscripts.

A list of some of the key early versions to consider include:

Latin Versions: Vetus Latina (Old Latin) is the earliest Latin translation, created before the Vulgate.[7] The translators used Greek manuscripts to translate the Latin versions and used them in different parts of the Roman Empire. Each manuscript shows variations, as there was no standard Latin version. We have over 10,000 fragments and complete manuscripts, though many are fragmentary or comprise only a few verses.[8] Pope Damasus I commissioned the **Vulgate** in the late 4th century, principally Jerome's work[9]; the Vulgate became the Catholic Church's officially promulgated Latin version of the Bible. We have thousands of copies, with over 8,000 manuscripts cataloged, ranging from complete Bibles to smaller fragments.[10]

Syriac Versions: Old Syriac includes the Sinaiticus and Curetonianus manuscripts,[11] dating from the 4th and 5th centuries. They are among the earliest Syriac translations and are essential for textual criticism. The Sinaitic Syriac and Curetonian Gospels are among the few known manuscripts.

Peshitta:[12] is a manuscript dating from the 5th century; it is a standard version of the Bible for churches in the Syriac tradition. There are hundreds of manuscripts, but the exact number is still being determined.

---

[7] Ibid 4.
[8] Berger, S. (2001). *The Old-Latin Gospels: A Study of their Texts and Language*. Oxford University Press.
[9] T. Lewis, Charlton; Short, Charles. "*A Latin Dictionary | vulgo*". www.perseus.tufts.edu. Retrieved 5 October 2019.
[10] Houghton, H. A. G. (2016). *The Latin New Testament: A Guide to its Early History, Texts, and Manuscripts*. Oxford University Press.
[11] Cureton, *Remains of a Very Ancient Recension of the Four Gospels in Syriac*, Hitherto Unknown in Europe, London, 1858; Cureton included an English translation of the newly discovered text, and a long introduction.
[12] Sebastian P. Brock. *The Bible in the Syriac Tradition* St. Ephrem Ecumenical Research Institute, 1988. Quote Page 13: "The Peshitta Old Testament was translated directly from the original Hebrew text, and the Peshitta New Testament directly from the original Greek"

Coptic Versions: translations of the New Testament made in Egypt and are significant for understanding early Egyptian Christianity. They include Sahidic Coptic, the earliest and most crucial Coptic version in Upper Egypt. We have several hundred manuscripts, including both complete books and fragments. Bohairic Coptic manuscripts, used in Lower Egypt, including the region around Alexandria, have fewer in number than Sahidic manuscripts, but still several hundred exist.[13]

Gothic Version: Created by Bishop Ulfilas in the 4th century, this is one of the earliest Germanic language translations of the Bible. We have a limited number of manuscripts, the most famous being the Codex Argenteus.[14]

Ethiopic Version: The Ethiopian translation of the New Testament is part of the broader Ge'ez Bible and is essential for the Ethiopian Orthodox Church. Several hundred manuscripts survived, though the exact count is hard to determine because of the dispersed nature of these manuscripts in collections worldwide.[15]

Armenian Version: The Armenian Bible, translated in the 5th century, is considered one of the earliest and most historically essential translations. Over 1,000 manuscripts exist, with many being kept in the Matenadaran in Yerevan, Armenia.[16]

Georgian Version: The manuscripts dating back to the 5th century reveal the Georgian translations of the New Testament. Scholars are aware of several dozen manuscripts, although the exact number must be well-documented.[17]

Each of these versions contributes uniquely to our understanding of the New Testament's textual history and the spread of early Christianity. The reason they are studied is not only for their theological significance but also for their linguistic, cultural, and historical value.

Quality of Manuscripts
Examining the consistency and integrity of the text across various manuscripts is a crucial aspect of biblical scholarship when assessing the accuracy in transmitting New Testament manuscripts. What sets the textual transmission of the New Testament apart from other ancient texts is its remarkable precision and the abundance of available manuscripts.

---

[13] Hyvernat, E. (1914). *Coptic Versions of the Bible. In The Catholic Encyclopedia.* New York: The Encyclopedia Press. http://www.newadvent.org/cathen/16078c.htm
[14] Metzger, B. M., & Ehrman, B. D. (2005). *The Text of the New Testament: Its Transmission, Corruption, and Restoration.* Oxford University Press.
[15] Burke, T. (2015). *Manuscripts of the Ethiopic New Testament.* In The Textual History of the Greek New Testament: Changing Views in Contemporary Research (pp. 437-452). SBL Press.
[16] Stone, M. E. (2014*). Armenian Versions. In The Textual History of the Greek New Testament: Changing Views in Contemporary Research* (pp. 453-468). SBL Press.
[17] Ibid 12.

Chapter 4 – Did Jesus Die? – Examining the Historical Evidence

Scholars have accounted for the entire New Testament text in manuscript form within 300 years of the original writing. Significant manuscript collections evidence this such as the Chester Beatty Collection, Bodmer Collection, Codex Sinaiticus, and Codex Vaticanus. This early dating of manuscripts is crucial, as for many other ancient Greek writings, like those of Plato or Aristotle, the earliest copies we have are often more than 1,000 years removed from the originals. Regarding manuscript support, the New Testament far surpasses these works, with only the Iliad by Homer coming close regarding manuscript numbers.[18]

Additionally, the writings of early Church Fathers are another vital source for studying the New Testament text. The writings of the Church Fathers alone, according to scholars like Metzger and Ehrman, could reconstruct almost the entire New Testament. These writings include numerous quotations from the New Testament and thus serve as an independent witness to its text.

As for the manuscripts themselves, there are different textual families or "types," with most Greek manuscripts belonging to the Byzantine textual family. Despite the many manuscripts and their early dates, variants exist in the textual transmission. However, these variants are often minor and do not significantly alter the meaning of the texts. These variants are unintentional, resulting from scribe errors like fatigue, faulty memory, or judgment. While there are intentional changes as well, these are generally not heretical or destructive but are more often attempts by scribes to improve the text in various ways, such as smoothing out grammatical or stylistic harshness.

The high degree of accuracy in the New Testament manuscripts, combined with their early dating and the corroborating witness of the Church Fathers, underscores the New Testament's strong bibliographical standing in classical antiquity. This body of evidence contributes to the overall reliability and integrity of the New Testament text as we have it today.[19]

Variations among New Testament manuscripts are a well-documented and crucial aspect of biblical textual criticism. These variations arise from various factors, including unintentional errors by scribes, differences in regional textual traditions, and intentional alterations. Unintentional Errors are the most common variants in manuscripts and typically result from the mechanical challenges of copying long texts by hand. They include errors like haplography, where a scribe might accidentally skip content, and parablepsis, when a scribe's eyes inadvertently jump from one part of the text to another, leading to omissions or duplications. These errors often produce grammatically incorrect readings or readings that need clarification within their context. Other common mistakes include orthographical

---

[18] Religious Studies Center. (n.d.). *New Testament Manuscripts, Textual Families, and Variants*. Retrieved from https://rsc.byu.edu/new-testament-manuscripts-textual-families-and-variants

[19] Defending Inerrancy. (n.d.). *Were the New Testament Manuscripts Copied Accurately?* Retrieved from https://www.defendinginerrancy.com/were-the-new-testament-manuscripts-copied-accurately/

differences or variations in spelling.[20]

The New Testament spread through various geographical regions, with each region potentially developing its own textual tradition. Scribes in different geographical regions could be influenced by these traditions, causing variations in the copying of manuscripts. The primary textual families scholars recognize include the Byzantine, Alexandrian, Western, and possibly Caesarean. Each family has distinct characteristics, such as the Byzantine text type's tendency to harmonize and conflate variant readings.

Sometimes, scribes make deliberate edits to the text to smooth out discrepancies or inaccuracies, update geographical information, harmonize texts based on familiar versions, or add details. Doctrinal debates or the desire to clarify ambiguous passages could also motivate these intentional changes.

Despite the variations, it's important to note that a vast majority of these differences are minor and do not significantly impact the New Testament's overall message or doctrinal content. The discipline of New Testament Textual Criticism involves comparing these manuscripts and using scientific methods to ascertain the most likely original text. The extensive number of manuscripts (about 5,800 Greek manuscripts) allows for careful comparison and a robust understanding of the textual form of the New Testament.[21]

Textual criticism is a complex and multifaceted process that involves various techniques to identify errors or differences in New Testament manuscripts and reconstruct the original text. The process begins with identifying variant readings by comparing the text of one manuscript with others. These variants can include additions, omissions, substitutions of words, or differences in the order of words. Scholars determine the origin, sequence, and relationships of these variants by categorizing and analyzing them.

Scholars have classified manuscripts into text types, such as the 'Western,' 'Alexandrian,' and 'Byzantine,' based on their lineage and geographical origins. The reliability of the Alexandrian text type is often attributed to its older manuscripts and the usage of its readings by early Church Fathers. Critical editions like the Novum Testamentum Graece have heavily influenced modern textual criticism, providing a comprehensive critical apparatus listing known textual variants. This facilitates the comparison of texts and aids in establishing a text that is as close as possible to the original.[22]

Textual critics evaluate the readings of various manuscripts, often favoring the Alexandrian text type, while some scholars advocate for the Western or Byzantine text types. The argument for the Byzantine text type is

---

[20] Ibid 18.
[21] The Center for the Study of New Testament Manuscripts. (n.d.). *Manuscripts 101: What is a textual variant?* Retrieved from https://www.csntm.org/manuscripts-101-what-is-a-textual-variant
[22] Parker, D. C. (2008). *Textual criticism. In An Introduction to the New Testament Manuscripts and their Texts.* Cambridge University Press. DOI: 10.1017/CBO9780511619922.006. Retrieved from Cambridge.org

based on the more significant number of manuscripts from later centuries, which indicates a closer alignment with the original texts.

The process of textual criticism involves carefully examining manuscripts, understanding their origins and relationships, and making informed judgments about the most likely original readings. This work is crucial for reconstructing the New Testament text as accurately as possible, considering the vast number of manuscripts and the variations they contain.

Scholars have a high degree of confidence in the current text of the New Testament, thanks to extensive work in textual criticism and a significant amount of manuscript evidence. Textual criticism is a rigorous academic process that is used not only for religious texts but for all ancient documents. It involves reconstructing a missing original from existing manuscripts, even when they are generations removed from the original document.

The reliability of the New Testament text is based on two key factors: the number of existing copies for comparison and the temporal proximity of these copies to the original writings. Compared to ancient secular texts, the New Testament has an impressive amount of manuscript evidence, with over 5,300 Greek manuscripts and numerous translations into other languages and quotations in early Christian writings. This is far more than any other ancient work.

Moreover, the earliest New Testament fragments date back to a very short period after the original compositions, enhancing confidence in their reliability. For example, scholars have dated the John Rylands Papyri, which contains a portion of the Gospel of John, around A.D. 117-138, indicating that the Gospel of John was circulating as far away as Egypt within 30 years of its composition.

Textual criticism is a process that employs established rules and methodologies to determine the extent of possible corruption or alteration of any work, including the New Testament. While there are about 300,000 individual variations of the New Testament text, most of these differences are minor, such as spelling errors or inverted phrases, and do not affect the overall message. When comparing the two leading text families of the New Testament, they show agreement about 98% of the time.[23]

Some of the earliest Church Fathers used these manuscripts, which are among the oldest, contributing to the general regard of textual critics for the Alexandrian text-type as the closest to the original autographs. However, there are different views, with some scholars advocating for the Western or Byzantine text types.

In summary, the extensive manuscript evidence, the close dating of the earliest manuscripts to the originals, and the rigorous work of textual criticism all contribute to a high degree of confidence in the reliability of the current New Testament text. While absolute certainty about every single word is impossible, the consensus is that the New Testament texts we have today are a reliable representation of the originals.

---

[23] Ibid 14.

Time Gap

The New Testament, particularly its Gospels, has a relatively short time gap between the original composition and the earliest surviving copies compared to many other ancient texts. This is a significant aspect in historical and textual analysis as it helps maintain a higher level of textual integrity and provides a strong foundation for textual criticism and historical analysis. Scholars can date the earliest known fragments of the New Testament to the 2nd century, such as Papyrus P52, which contains a portion of the Gospel of John and dates back to around 125-175 AD. This dating is relatively close to the estimated original composition date of John's Gospel, which is late 1st century to early 2nd century AD.[24]

By the 3rd and 4th centuries, more substantial portions of New Testament texts were available, such as Vaticanus and Sinaiticus Codex from the 4th century, which contain nearly complete compilations of the New Testament. Compared to other ancient texts, where gaps of several centuries are common, the New Testament's textual history is more robust. This allows for a more confident reconstruction of the original text. Even though textual criticism of the New Testament reveals variations among manuscripts, most biblical scholars and textual critics agree that the overall integrity of the texts has been well-preserved, especially considering the time and manner in which scribes copied and transmitted them.[25]

The existence of early manuscripts does not prove the historical reliability of the events described in the texts, but it strengthens the case for considering the New Testament writings as critical historical sources from their time. The relatively short time gap between the composition and the earliest copies of the New Testament is a crucial factor in its study. It offers an opportunity to explore the historical and textual significance of the New Testament, and to better understand the context and background of the events and people described in its pages.[26]

To gain an understanding of the context in which we bibliographically analyze New Testament manuscripts, we must examine ancient writings and the number of manuscripts that have survived to the present day. For instance, Caesar's Gallic War has an 850-year gap, which makes it impossible to verify the original context. In addition, we can go even further back in time to when someone wrote Homer's Iliad around 800 BC, and they discovered the first copy 400 years later, giving us an estimated date of 415 BC.

We may wonder how many people would question these ancient works, but the gap in literary masterpieces from ancient times gives us a more realistic perspective when compared to the New Testament manuscripts. If we consider classical authors like Plutarch, Tacitus, Polybius, Thucydides,

---

[24] Wallace, Daniel B. "*The Gospel According to Bart: A Review Article of Misquoting Jesus by Bart Ehrman.*" Journal of the Evangelical Theological Society 49, no. 2 (2006): 327-49.
[25] Metzger, Bruce M. *The Text of the New Testament: Its Transmission, Corruption, and Restoration.* Oxford University Press, 2005.
[26] Aland, Kurt, and Barbara Aland. *The Text of the New Testament: An Introduction to the Critical Editions and to the Theory and Practice of Modern Textual Criticism.* Eerdmans, 1995.

or Xenophon, we find that less than 10 extant copies exist, and the earliest copies were written 750-1500 years after the original manuscript. Therefore, it is crucial to examine the reliability of surviving texts carefully.

ANCIENT MANUSCRIPT COMPARISON CHART

Author	Date Written	Earliest Copy	Approximate Time Span between original & copy	# of Copies	Accuracy of Copies
Lucretius	died 55 or 53 B.C.		1100 yrs	2	----
Pliny	A.D. 61-113	A.D. 850	750 yrs	7	----
Plato	427-347 B.C.	A.D. 900	1200 yrs	7	----
Demosthenes	4th Cent. B.C.	A.D. 1100	800 yrs	8	----
Herodotus	480-425 B.C.	A.D. 900	1300 yrs	8	----
Suetonius	A.D. 75-160	A.D. 950	800 yrs	8	----
Thucydides	460-400 B.C.	A.D. 900	1300 yrs	8	----
Euripides	480-406 B.C.	A.D. 1100	1300 yrs	9	----
Aristophanes	450-385 B.C.	A.D. 900	1200	10	----
Caesar	100-44 B.C.	A.D. 900	1000	10	----
Livy	59 BC-AD 17	----	???	20	----
Tacitus	circa A.D. 100	A.D. 1100	1000 yrs	20	----
Aristotle	384-322 B.C.	A.D. 1100	1400	49	----
Sophocles	496-406 B.C.	A.D. 1000	1400 yrs	193	----
Homer (Iliad)	900 B.C.	400 B.C.	500 yrs	643	95%
New Testament	1st Cent. A.D. (A.D. 50-100)	2nd Cent. A.D. (c. A.D. 130 f.)	less than 100 years	5600	99.5%

NOTES:
- There are thousands more New Testament Greek manuscripts than any other ancient writing.
- The internal consistency of the New Testament documents is about 99.5% textually pure.
- In addition, there are over 19,000 copies in the Syriac, Latin, Coptic, and Aramaic languages. The total supporting New Testament manuscript base is over 24,000.

SOURCE: Christian Apologetics and Resource Ministry - https://carm.org/manuscript-evidence

Scholarly Consensus

The Gospels in the New Testament have been the subject of much scholarly debate and discussion over the years. While biblical scholars and textual critics broadly agree that the Gospels we possess today closely resemble the original writings in terms of content, much of the scholarly debate surrounding them focuses on their interpretation, meaning, and historical context.

The abundance of manuscript evidence and centuries of scholarly examination have contributed to the New Testament, including the Gospels, being considered well-preserved in terms of its textual integrity. While textual variants exist among the manuscripts, most differences are minor and do not significantly alter the fundamental content or message of the texts.[27]

Scholars often employ a historical-critical approach to understand the

---

[27] Evans, C.A. (2012). *The Reliability of the New Testament: Bart Ehrman and Daniel Wallace in Dialogue*. Fortress Press.

Gospels' context, authorship, intended audience, and historical accuracy. This approach leads to diverse understandings and interpretations of the text, reflecting different theological, denominational, and academic perspectives.[28]

It's important to distinguish between the textual reliability of the Gospels and the historical reliability of the events they describe. Questions about the historicity of specific events, miracles, or the accuracy of theological claims are separate from the issue of textual reliability.

Despite the scholarly consensus that the Gospels in the New Testament are textually reliable and close to their original composition in content, significant debate and discussion continue regarding these texts' interpretation, meaning, and historical context. This reflects the diverse and dynamic nature of biblical scholarship, and the ongoing quest to understand the texts within their cultural and historical contexts.[29]

In summary, the Bibliographical Test suggests that the Gospels, as part of the Christian scriptures, were passed down through generations with significant precision and faithfulness. This aspect is crucial when evaluating their credibility as documents of historical significance. It's important to note that while this test demonstrates the consistent transmission of these texts over time, it does not inherently validate or verify the factual accuracy of the specific events and narratives they contain. The test focuses more on textual integrity and preservation than the empirical truth of the incidents recounted in these writings.

Internal Evidence for the Gospels

The internal evidence supporting the historicity of Jesus' death comes primarily from the New Testament documents. Intrinsic evidence supporting the historicity of Jesus' death is found primarily within the biblical texts, rather than relying on external confirmations. There are several different types of internal evidence that contribute to our understanding of the historical accuracy of the biblical accounts of Jesus' death.

One type of internal evidence is literary characteristics. For example, the Gospels exhibit a narrative style that aligns with other ancient biographical and historical texts. The authors of the Gospels also make use of literary devices such as foreshadowing, allusion, and irony that suggest a high degree of literary skill and intentionality.

Another type of internal evidence is consistency among the Gospels. Despite being written by different authors at different times and for different audiences, the Gospels contain numerous details that are consistent with one another. This suggests that the authors were drawing on a shared body

---

[28] Burkett, D. (2010). *An Introduction to the New Testament and the Origins of Christianity*. Cambridge University Press.
[29] Metzger, B.M. (2005). *A Textual Commentary on the Greek New Testament*. Hendrickson Publishers.

of knowledge about the life and death of Jesus.

Contextual details are another type of internal evidence. The Gospels contain numerous references to people, places, and events that are consistent with what we know about the historical and cultural setting of the time. For example, the Gospels describe the political and religious tensions that existed in first-century Palestine, which are consistent with what we know about that period from other historical sources.

Finally, the portrayal of events in a manner consistent with the historical and cultural setting of the time is another type of internal evidence. For example, the Gospels describe Jesus' crucifixion in a way that is consistent with what we know about Roman crucifixion practices from other historical sources. Taken together, these different types of internal evidence provide valuable insights into the historicity of Jesus' death.

Consistency Among the Gospels:

Despite differences in specifics, the four Gospels – Matthew, Mark, Luke, and John – share a cohesive story about the arrest, trial, crucifixion, and demise of Jesus. This uniformity across separate sources implies a shared foundational historical occurrence. Matthew, Mark, Luke, and John – the four canonical Gospels, provide the main accounts of Jesus' demise in the New Testament. Authored in the first century AD, these texts not only form the core of Christian doctrine but also act as historical records to a certain degree.

The Gospels' accounts about the narrative of Jesus's death are as follows:

- **Matthew (Matthew 27:32-56)**: This Gospel describes Jesus being handed over to Roman soldiers to be crucified. It details the mocking and torture he endured, his crucifixion at a place called Golgotha, and the events during his crucifixion, such as the darkness over the land and the tearing of the temple curtain. It also mentions the presence of Mary Magdalene and other women witnessing the event.

- **Mark (Mark 15:21-41)**: Mark's account is similar to Matthew's, with specific details about Simon of Cyrene carrying the cross, the inscription of the charge against Jesus ("The King of the Jews"), and Jesus' last moments and death cry. It also notes the reaction of the Roman centurion who observed Jesus' death.

- **Luke (Luke 23:26-49)**: Luke includes unique elements like Jesus' words to the women who mourned him on the way to the crucifixion and his interaction with the criminals crucified alongside him. It emphasizes Jesus' forgiveness towards his executioners and final words, committing his spirit to God.

- **John (John 19:16-37)**: John's Gospel provides a more detailed account of the crucifixion, highlighting interactions between Jesus, his mother, and the beloved disciple. It includes specific details like the soldiers casting lots for Jesus' clothes and the piercing of Jesus' side by a Roman soldier.

Consistency in the essential storyline among the Gospels: Matthew, Mark, Luke, and John[30], despite originating from different authors and perspectives, exhibit a remarkable consistency in their portrayal of several key events in Jesus' life, especially concerning his arrest, trial, crucifixion, and death. This harmonious core narrative is significant for a few reasons. Despite being written independently, all the Gospels converge on these crucial aspects of Jesus' life.[31] Despite their individual stylistic and thematic differences, this convergence suggests a common historical foundation.

The authors of these Gospels came from varied backgrounds and wrote for different audiences. Matthew[32], for example, focuses on a Jewish audience, while Luke[33] addresses a more Gentile readership. This diversity in viewpoint adds layers of complexity and nuance to the narrative yet does not detract from the consistency of the critical events, thereby enhancing the overall credibility of these accounts.

The Gospels do not merely agree on the general outline of these events but also align in several specific details. For instance, they mention key figures like Pontius Pilate, the Roman governor, and Caiaphas, the high priest, playing pivotal roles in Jesus' trial and execution. Such specificities further strengthen the case for a joint historical basis. The detailed variations, far from undermining the narrative, are hallmarks of authentic independent testimonies. In real-life events, different witnesses often recall the same event with detailed variations due to their perspectives and biases. The Gospels demonstrate this natural diversity, which can be seen as corroborative rather than contradictory.

All four Gospels reflect an awareness of the social, religious, and political realities of 1st-century Judea, as they are embedded in its cultural and historical context. This con textuality supports the historical plausibility of the events they describe. The coherence and consistency among the Gospels regarding Jesus' arrest, trial, crucifixion, and death, balanced with the natural differences expected from independent sources, provide a solid basis for considering these accounts as rooted in a joint historical event. This does not necessarily confirm every detail of each account but does suggest that there is a significant historical core around which these narratives are constructed.

Historical and Cultural Context

The Gospels provide a nuanced portrayal of the intricate relationship

---

[30] Burkett, D. (2002). *The Gospel According to John.* Oxford University Press.
[31] Evans, C. A. (2012). *The Gospels and Acts.* Baker Academic.
[32] Keener, C. S. (2012). *The Gospel of Matthew: A Socio-Rhetorical Commentary.* Eerdmans.
[33] Green, J. B. (1997). *The Gospel of Luke.* Eerdmans.

between Jewish religious leaders and the Roman political authorities.[34] This complex dynamic is historically accurate for the period. The Sanhedrin, the supreme Jewish religious council, held significant sway over Jewish religious matters but operated under the overarching authority of the Roman Empire. In the portrayal: This delicate balance of power is depicted in the Gospels through the involvement of Jewish and Roman authorities in Jesus' trial.

The involvement of Jewish leaders, particularly the Sanhedrin, in Jesus' trial is detailed in the Gospels. This is consistent with the historical role of the Sanhedrin, which was responsible for religious and civil matters within the Jewish community. However, the Sanhedrin's jurisdiction was limited by Roman oversight, especially when it came to capital punishment cases. The Gospels' depiction of Pontius Pilate, the Roman prefect of Judea, is consistent with historical records. Historical records reveal that Pilate was known for his often-brutal rule and possessed the final authority in legal matters, including the power to order executions. The Gospels' depiction of Pilate's involvement in Jesus' trial is consistent with his role as the ultimate judicial authority in the region.[35]

The Gospels reflect the socio-political tensions of the time. Judea was a region under Roman occupation, and there were constant tensions between the Jewish population and the Roman authorities. The Gospels capture this backdrop effectively, providing context for the events leading up to Jesus' crucifixion. We also see an accurate representation of the time's various Jewish religious and cultural practices. This includes references to festivals, religious rituals, and the Pharisees and Sadducees, different sects within Judaism with distinct beliefs and practices. Such details add to the historical credibility of the narratives.

The geographical and topographical details in the Gospels align well with the known information about the region during that time. The Gospels provide an accurate depiction of locations such as Jerusalem, the Sea of Galilee, specific temples, and roads, further grounding the narratives in a natural historical setting. The language used in the Gospels, including idiomatic expressions and titles, is appropriate for the period and setting. This linguistic accuracy provides an additional layer of historical authenticity.[36]

The Gospels' portrayal of the role of Jewish and Roman authorities and the broader socio-political and cultural context of the time aligns well with historical records and archaeological findings. This accuracy in depicting the complex dynamics and the cultural and religious milieu of 1st-century Judea lends credibility to their narratives. It situates them firmly within the historical context of the era.

---

[34] Crossan, J. D. and Reed, J. L. (2004). *Excavating Jesus: Beneath the Stones, Behind the Texts.* New York: HarperCollins.

[35] Evans, C. A. (2012). *The Historical Christ and the Jesus of Faith: The Incarnational Narrative as History.* Oxford: Oxford University Press.

[36] Hengel, M. (1977). *The Son of God: The Origin of Christology and the History of Jewish-Hellenistic Religion.* Philadelphia: Fortress Press.

Crucifixion Practices

The Gospels' description of crucifixion adheres closely to what historical and archaeological evidence suggests about Roman execution methods. Crucifixion was a common form of execution used by the Romans, particularly for enslaved people, revolutionaries, and the most heinous criminals. It was designed to be a slow, painful, and publicly humiliating way to die. The Gospels mention Jesus carrying the cross beam to the crucifixion site. Historically, it was customary for the condemned to carry the patibulum (cross beam) to the place of execution. Typically, the execution site would already have the vertical beam, or stipes, erected. This practice not only added to the physical suffering but also served as a part of the humiliation, with the condemned paraded through the streets.[37]

Accounts of Jesus being nailed through his hands and feet are consistent with Roman crucifixion methods. Archaeological findings, such as the remains of crucified individuals with nail marks, confirm that Roman crucifixion methods commonly involved nailing, although some victims were also tied to the cross. This method was intended to prolong the agony of the victim. The Gospels' depiction of guards at the crucifixion site matches Roman practices. To prevent any attempts at rescuing the condemned and to ensure that the sentence was carried out, guards were often stationed by the Romans. Their presence also served as a deterrent to potential rebels or criminals.

The Gospels describe Jesus' crucifixion taking place outside the city walls, which aligns with historical practices. Romans typically carried out executions in prominent locations outside city limits, both for sanitary reasons and to maximize the public spectacle, serving as a warning to others. Considering the timing and public nature, it is historically plausible that the crucifixion occurred during a major festival (Passover). Executions were frequently utilized by the Romans during such occasions to make a powerful public statement, particularly when addressing alleged insurrectionists or individuals accused of crimes against the state.

Consistent with Roman practice, the mention of the titulus is made, which is the inscription above Jesus' head stating his alleged crime. Such inscriptions were used to state the crime and serve as a warning to others. The Gospels mention soldiers breaking the legs of the two individuals crucified alongside Jesus and offering a sedative (wine mixed with myrrh) to Jesus. These details correspond with known Roman practices. Breaking the legs hastened death by asphyxiation, and offering sedatives was a form of minor mercy, sometimes practiced to lessen the agony of the crucified.

The detailed portrayal of crucifixion in the Gospels aligns well with the historical and archaeological understanding of Roman execution methods. This alignment adds to the historical authenticity of the accounts, providing a vivid and accurate depiction of what was a brutal and significant aspect of

---

[37] Biblical Archaeology Society. (n.d.). *Roman Crucifixion Methods Reveal the History of Crucifixion*. Retrieved from https://www.biblicalarchaeology.org/daily/biblical-topics/crucifixion/roman-crucifixion-methods-reveal-the-history-of-crucifixion/

Roman justice at the time.

Literary Style and Structure

Specific, concrete details characteristic of eyewitness testimony mark the Gospels. For example, the mention of particular names (like Simon of Cyrene, who helped carry the cross), specific places (such as the Garden of Gethsemane), and small, vivid details (like the soldiers casting lots for Jesus' garments) suggest a narrative style that stems from direct observation. These are not the kind of generic descriptions one would expect from second-hand stories or later fabrications.[38]

The Gospels often include personal reactions and emotions indicative of eyewitness accounts. Descriptions of the disciples' fear, Peter's denial, or the women's sorrow at the crucifix provide a personalized touch consistent with accounts from individuals who directly witnessed the events. The Gospels exhibit consistency in narrating critical events while differing in some details. This is typical of multiple eyewitness accounts where each witness observes and recalls the same event slightly differently based on their perspective, focus, and psychological state. Such variations lend credibility to the accounts, reflecting the natural differences in human observation and memory.[39]

The Gospels include details that do not necessarily advance the narrative but add to the authenticity of the accounts. For example, the mention of the number of fish caught in John 21:11 or the color of the robe placed on Jesus during his trial. These details resemble the kind of incidental memories an eyewitness might retain. The dialogues and interactions among characters in the Gospels bear the marks of realism. Conversations, especially those involving Jesus, often reflect the spontaneity and complexity of natural speech rather than being stylized or overly formal.

The Gospels sometimes reflect a particular perspective or point of view, suggesting an eyewitness basis. For instance, the Gospel of John contains details (like the disciple Jesus loved) that imply a personal presence and perspective. Eyewitness accounts often include sensory information – what they saw, heard, or felt. The Gospels contain descriptions, like the darkness that fell over the land, the earthquake at the time of Jesus' death, or the description of physical appearances and sounds.

The Gospels often describe immediate reactions to events, such as the disciples' dismay at Jesus' arrest or the women's initial response to the empty tomb. These emotional and spontaneous reactions are typical of eyewitness descriptions, where the emotional impact is profound and directly conveyed. The literary style and structure of the Gospels, with their specific details, personalization, realistic dialogue, and sensory descriptions, are indicative of eyewitness accounts. These elements contribute to the authenticity and vividness of the narratives, suggesting that the authors were

---

[38] Dibelius, M. (1927). *The Structure and Literary Character of the Gospels*. The Harvard Theological Review, 20(3), 151–170. http://www.jstor.org/stable/1507902

[39] Faraoanu, E. (n.d.). *Title of the Document*. Retrieved from http://www.ejst.tuiasi.ro/Files/74/18_Faraoanu.pdf

either direct witnesses to the events they describe or had access to first-hand testimonies.

Continuing looking at the literary style and structure, we can observe the use of Aramaic phrases in the Gospels. The Gospels, primarily written in Greek, notably include Aramaic phrases, particularly in critical moments such as Jesus' words on the cross. This inclusion is significant because Aramaic was the common language spoken in Judea during the 1st century, whereas Greek was the lingua franca of the broader Hellenistic world. The preservation of these Aramaic phrases within a Greek text suggests an effort to maintain the authenticity and immediacy of the original sayings.[40]

The use of Aramaic phrases points to the historical and cultural context in which the events of the Gospels occurred. By retaining these phrases, the authors provide a more authentic snapshot of the linguistic landscape of Judea at the time, enhancing the historical credibility of the accounts. The Aramaic terms used in the Gospels often carry significant emotional or theological weight. For instance, Jesus' cry from the cross, "*Eli, Eli, lema sabachthani?*" (which translates to "My God, my God, why have you forsaken me?"), is a direct quote in Aramaic. Its retention in the original language adds to the dramatic and emotional intensity of the moment, suggesting a direct connection to the historical Jesus.

Including Aramaic phrases might also indicate that the sources or eyewitnesses behind these Gospel accounts were native Aramaic speakers. This lends credence to the theory that the Gospels, or at least the sources they were based on, originated from within the community that personally knew or heard Jesus. In an oral culture, memorable sayings or phrases in a familiar language could aid memory. The preserved Aramaic phrases in the Gospels might have served as crucial mnemonic tools, aiding the early Christian community in remembering and faithfully passing on the teachings and events of Jesus' life.[41]

By translating most of the Gospels into Greek while preserving specific Aramaic phrases, there was an intention to make the teachings of Jesus accessible to a wider Hellenistic audience without compromising the essence and authenticity of his original words. The worship and collective memory of the early Christian community likely held a special place for the Aramaic phrases in the Gospels. Their preservation in the texts would have been a powerful link to the historical Jesus and the origins of their faith. Aramaic phrases within the predominantly Greek Gospels provide a compelling layer of authenticity to the narratives. It reflects the linguistic reality of the time, suggests a close connection to eyewitness sources, and preserves the emotional and theological impact of the original words of the historical figures, particularly of Jesus himself.

---

[40] Roberts, M. D. (2010). *Jesus and Aramaic in the Gospels*. Retrieved from https://www.beliefnet.com/columnists/markdroberts/2010/07/jesus-and-aramaic-in-the-gospels.html
[41] Bryn Mawr Classical Review. (n.d.). *Aramaic Sources of Mark's Gospel*. Society for New Testament Studies Monograph Series, 102. Retrieved from https://bmcr.brynmawr.edu/1999/1999.04.21/

Chapter 4 – Did Jesus Die? – Examining the Historical Evidence

Fulfillment of Old Testament Prophecies

For many believers, one of the compelling aspects of the Gospels is how Jesus' life and death fulfill various prophecies found in the Old Testament, particularly in books like Isaiah and Psalms. Believers interpret these fulfillments not as coincidental but as a manifestation of a divine plan unfolding through history. Prophecies such as the suffering servant in Isaiah 53, the righteous but afflicted individual in Psalms 22, and the specific details about birth, ministry, death, and resurrection directly point to Jesus.[42] Believers often interpret these connections as internal evidence that validates the historical reliability of the Gospel narratives and affirms the divine nature of Jesus' mission.

Specific examples often cited include:

- The nativity narratives interpret the prophecy of a virgin birth (Isaiah 7:14) as being fulfilled.
- The fulfillment of the prediction that the Messiah would be a descendant of David is seen in the genealogies of Jesus.
- The fulfillment of the foretelling of the Messiah's suffering and death for humanity's sins is evident in the accounts of Jesus' crucifixion.

Critics and some scholars, on the other hand, approach these perceived fulfillments more skeptically. They often argue that the Gospel writers, aware of the Old Testament prophecies, could have shaped their narratives to reflect these predictions retrospectively. From this perspective, the fulfillment of prophecies in the Gospels is seen not as evidence of a divine plan but as a literary and theological construct by the early Christian community. This interpretation suggests that the Gospel writers, seeking to establish Jesus' messianic credentials, may have intentionally included or emphasized elements in their narratives that mirrored specific Old Testament prophecies.

Some scholars note that the methodology employed by the Gospel writers was typical of religious and historical writing of the period. Ancient writers often sought to link new events or figures to revered traditions or texts, in this case, the Hebrew Scriptures, to provide legitimacy and context. The Gospel writers, therefore, might have been employing a common literary technique to connect Jesus with Old Testament prophecy, reinforcing his significance in the eyes of their contemporaries.

The discussion of prophecy fulfillment has profound theological implications; it strengthens faith and the divine authority of the scriptures. For critics, it raises questions about the Gospels' historical accuracy and the authors' intent. This debate underscores the complexity of interpreting ancient religious texts and the intersection of faith, history, and theology.

They understand the cultural and historical context of the Old Testament

---

[42] Nicholas J. Frederick, "*The Use of the Old Testament in the New Testament Gospels*," in Prophets and Prophecies of the Old Testament, ed. Aaron P. Schade, Brian M. Hauglid, and Kerry Muhlestein (Provo, UT: Religious Studies Center; Salt Lake City: Deseret Book, 2017), 123-160.

prophecies and the New Testament writings. To understand the role of prophecy fulfillment in the Gospels, it is crucial to comprehend how prophecies were understood and interpreted in the Jewish tradition of the time, and how these interpretations influenced early Christian thought. The fulfillment of Old Testament prophecies in the life and death of Jesus, as presented in the Gospels, remains a topic of significant debate. Believers see it as evidence of divine orchestration and historical reliability, while critics often view it as a case of retrospective interpretation. This debate highlights the complexities of interpreting religious texts and the interplay between faith and historical analysis.

Early Christian Testimony

The letters of Paul, also known as the Pauline Epistles, hold a crucial place in early Christian testimony due to their chronology. These epistles provide some of the earliest written accounts of Christian beliefs and traditions before the Gospels. The fact that they come before the Gospel narratives is significant. Paul's letters reference critical events in the life of Jesus, particularly his death and resurrection, and the significance of these events in Christian theology. While the epistles do not provide a detailed narrative of Jesus' life as the Gospels do, the elements Paul discusses are consistent with the later Gospel narratives. For instance, Paul's emphasis on Jesus' death and resurrection aligns closely with the central themes of the Gospels.[43]

One reason Paul's writings are so significant is his independent source of information. Paul was not one of the original twelve apostles of Jesus; he converted to Christianity after Jesus' death and resurrection. His teachings and understandings are based on his revelations and interactions with the early Christian community rather than direct contact with Jesus during his earthly ministry. This independence makes his writings a valuable, separate testimony about early Christian beliefs.

Paul's letters provide a theological framework that underpins much of Christian doctrine. Paul's letters establish a theological framework that supports much of Christian doctrine. His discussions of concepts like salvation, grace, faith, and the role of Jesus as the Messiah form the foundation of Christian theology and find resonance in the Gospel narratives. Paul's interpretation of Jesus' life and teachings significantly shaped early Christian thought and practice. Paul's epistles also offer insights into the historical and sociocultural context of the early Christian community. His letters address various issues, challenges, and controversies facing early Christians, reflecting the dynamics of the community as it expanded beyond its Jewish roots into the broader Greco-Roman world.

Many scholars widely accept the authenticity of Paul's letters, making them some of the most reliable sources for understanding early

---

[43] Turner, R. (2009, October 1). *An Analysis of the Pre-Pauline Creed in 1 Corinthians 15:1-11*. Retrieved from https://carm.org/about-pre-pauline-creed-in-1-corinthians-15

Christianity.[44] Unlike later texts, these epistles are less influenced by the developing Christian tradition and more reflective of the immediate post-Jesus period. We cannot overstate the influence of Paul's writings on early Christian thought. His interpretation of Jesus' teachings and his theological expositions were instrumental in forming early Christian doctrine and practice, influencing the content and interpretation of the Gospels.

Pauline Epistles is a critical source of early Christian testimony. Written earlier than the Gospels, they offer a consistent, albeit less narrative-focused, account of Jesus' death and its significance. They also provide a separate, independent framework that complements and corroborates the Gospel narratives. Paul's writings are significant not only for their chronological precedence but also for their independent origin and substantial influence on the development of early Christian theology and doctrine.

Conclusion

Supporting the proof of Jesus' death, the internal evidence found in the Gospels has played a crucial role. Jesus' trial, crucifixion, and burial are extensively documented in the Gospels, offering a detailed account of his last moments. The belief that Jesus indeed died is strengthened by the internal evidence in the Gospels, just as the Gospels have suggested. Scholars and historians alike have questioned the validity of the historical evidence found in the Gospels, subjecting it to intense scrutiny over the centuries. However, when analyzed through the lens of internal evidence, the proof of Jesus' death becomes more apparent.

Internal evidence refers to the details that are present within the text itself. In the case of the Gospels, these details include the eyewitness accounts of the crucifixion, the reactions of those present at the time, and the details of Jesus' burial. External sources cannot provide the same level of authenticity as these details do. Additionally, external sources have provided corroboration for the internal evidence found in the Gospels. For example, the Roman historian Tacitus wrote about the crucifixion of Jesus in his Annals, providing further evidence of its occurrence.

In conclusion, the internal evidence found in the Gospels provides a compelling case for the proof of Jesus' death. While external sources have also contributed to the evidence, the internal details found within the text itself establish a level of authenticity that cannot be found elsewhere.

---

[44] While seven of, 1972), 124-28. the letters traditionally attributed to Paul (Romans, 1-2 Corinthians, Galatians, Philippians, 1 Thessalonians, and Philemon) are routinely accepted as authentic in modern scholarship, Ephesians, Colossians, 2 Thessalonians, 1-2 Timothy, and Titus remain disputed. For a relatively recent overview, see MacDonald, Margaret T. "The Deutero-Pauline Letters in Contemporary Research," in The Oxford Handbook of Pauline Studies (New York: Oxford University Press, 2014), 258-279.

External Evidence for the Gospel Narratives

Historical Writings from Non-Christian Sources

One of the primary forms of external evidence for the Gospels comes from historical accounts written by non-Christian authors of the time. Roman historians, such as Tacitus,[45] provide a significant non-Christian perspective on early Christianity in his "Annals," written around 116 AD. In discussing the Great Fire of Rome and Nero's subsequent persecution of Christians, Tacitus mentions Jesus in the context of his execution by Pontius Pilate during the reign of Tiberius. This reference is crucial for several reasons:

- It confirms the historical existence of Jesus independent of Christian sources.
- Tacitus' mention of Pontius Pilate as the prefect of Judea during Jesus' execution aligns with the Gospel accounts.
- Tacitus' account carries weight in historical scholarship as a Roman historian known for his attention to detail and reliability.

Another author of this period is Flavius Josephus,[46] a Jewish historian who wrote "Antiquities of the Jews" in the late first century AD. This work contains a passage known as the Testimonium Flavianum, which mentions Jesus, his teachings, his crucifixion under Pontius Pilate, and even his resurrection. This passage is significant but also controversial:

- Some scholars argue that later Christian writers may have added parts of this passage, as it contains language uncharacteristic of Josephus, who was not a Christian.
- Despite these concerns, most scholars agree the core reference to Jesus is likely authentic, suggesting that Josephus wrote about Jesus, albeit less extensively and without the Christianizing additions.
- The reference to Pontius Pilate and the crucifixion aligns with the Gospel narratives.

It is essential to know that Tacitus and Josephus are reliable historians of their time. Their references to Jesus and the early Christians add a layer of historical validation to the Gospel narratives from non-Christian sources. These accounts provide external corroboration for specific details in the Gospels, such as the existence of Jesus, his role as a teacher or leader, and his crucifixion under Pontius Pilate. The fact that these references come from authors outside the Christian tradition is crucial, as it reduces the likelihood of bias favoring Christian theological positions.

While these sources corroborate some aspects of the Gospel narratives, they do not provide detailed accounts of Jesus' life or teachings. The

---

[45] Tacitus. (c. 116 AD). *The Annals*. Book 15, Chapter 44.
[46] Flavius Josephus. (c. 94 AD). *Antiquities of the Jews*. Book 18, Chapter 3.

Chapter 4 – Did Jesus Die? – Examining the Historical Evidence

motivations of these authors differed from those of the Gospel writers. Tacitus and Josephus wrote from a historical perspective, focusing on the broader context in which Christianity emerged rather than on the religious significance of Jesus' life.

Understanding the context in which Tacitus and Josephus wrote is essential for interpreting their accounts. Tacitus wrote with a typical Roman disdain for Christianity, while Josephus, a Jew who worked under Roman patronage, had a complex relationship with both Judaism and the Roman authorities. Their works reflect the perceptions and attitudes of their respective cultures towards early Christians, providing valuable insights into the historical and social backdrop of the early Christian era.

Historical writings from non-Christian sources like Tacitus and Josephus play a vital role in the external validation of certain aspects of the Gospel narratives. They provide independent attestations of some critical elements, such as the historical existence of Jesus and his crucifixion under Pontius Pilate. However, one should consider their accounts within the broader historical context and understand that these sources offer perspectives distinct from the Gospels' theological focus.

Archaeological Discoveries

The Pilate Stone,[47] discovered in 1961 at Caesarea Maritima in Israel, is a significant archaeological find substantiating a crucial figure mentioned in the New Testament, Pontius Pilate, who served as the Roman prefect of Judea. This limestone block with the Latin inscription "Pontius Pilatus, Prefect of Judea" provides concrete evidence of Pilate's historical existence and administrative role as stated in the Bible. As one of the few archaeological artifacts directly connecting to a specific individual mentioned in the Gospels, the stone is notable for offering tangible evidence for the historical context of the narratives.

Unearthed during an excavation in the ancient Roman city of Caesarea Maritima, the Pilate Stone has been a significant subject of study in the field of Biblical archaeology. Corroborating Biblical accounts, particularly those involving Pontius Pilate in the trial and crucifixion of Jesus Christ, the inscription on the stone holds great significance. Its discovery marked a milestone in Biblical archaeology, confirming Pilate's role in Judea and providing a non-Biblical corroboration of his governorship. The Israel Museum in Jerusalem currently houses the stone, while plaster-cast replicas are showcased in various locations, including Caesarea Maritima.

This finding enriches our understanding of the historical context of the New Testament, serving as a bridge between religious texts and historical evidence. It also provides insights into the Roman administrative system in Judea during the first century CE. The Pilate Stone's contribution to Biblical scholarship is profound, as it supports the historical accuracy of the New Testament, particularly in the accounts related to Pontius Pilate and the

---

[47] Wroe, Anne (April 3, 1999). *"Historical Notes: Pontius Pilate: a name set in stone"*. The Independent.

crucifixion of Jesus.

Archaeological findings in Jerusalem have provided substantial evidence supporting the historical accuracy of the New Testament. These discoveries encompass various aspects of the setting and context in which the events of the New Testament are narrated. Here is a detailed list of some significant archaeological findings:

1. **The Pool of Siloam**: In the Gospel of John, the Pool of Siloam is mentioned in relation to Jesus healing a blind man. Archaeologists have uncovered a first-century ritual pool near the mouth of Hezekiah's tunnel, confirming the existence and location of the Pool of Siloam as described in the New Testament.[48]

2. **Nazareth**: The hometown of Jesus, Nazareth,[49] was once considered insignificant and not mentioned in any surviving literature until after the time of Jesus. However, archaeological evidence now confirms the existence of Nazareth in the first century, aligning with the biblical narrative.

3. **Coastal Cities Along the Sea of Galilee**: Several cities mentioned in the Biblical narratives surround the Sea of Galilee, a significant location in the Gospels, particularly in the context of Jesus' ministry. Four of these cities – Capernaum, Bethsaida, Chorazin, and Tiberias – have been the focus of archaeological efforts to ascertain their historical locations and contexts.[50]
    - **Capernaum (Tell Hum):** Scholars identify the site of Tell Hum with Capernaum, a place mentioned frequently in the Gospels as a center of Jesus' activities in the region. Archaeological excavations at Tell Hum have uncovered structures that align with the period of Jesus' life, adding credence to the Gospel accounts.
    - **Bethsaida**: Located near the northern shore of the Sea of Galilee, Bethsaida is another city mentioned in the context of Jesus' ministry. Ongoing archaeological work aims to confirm its exact location, with several potential sites under consideration.
    - **Chorazin**: Although less prominent in the Gospels, Chorazin is mentioned in the context of Jesus' teachings. Archaeological discoveries in the area have yielded findings that are consistent with a first-century settlement, providing a tangible connection to the Gospel narratives.

---

[48] "*Archaeologists identify traces of 'miracle' pool*". NBC News. 2004-12-23. Retrieved 2023-10-29.

[49] Jeffrey, David L. (1992). *A Dictionary of biblical tradition in English literature*. Wm. B. Eerdmans Publishing. pp. 538–40. ISBN 978-0-85244-224-1. Archived from the original on 2020-10-08. Retrieved 2020-11-01.

[50] Reed, J. L. (2002). *Archaeology and the Galilean Jesus: A Re-examination of the Evidence*. Harrisburg, PA: Trinity Press International.

- **Tiberias**: Unlike the other cities, Tiberias is known to have been established later, during the Roman period. It's mentioned in historical sources and its location has been continuously inhabited, making it easier to verify in the context of Jesus' time.

4. **Crucifixion Victim**: A crucifixion victim named Yehohanan was discovered in a burial site in Giv'at ha-Mivtar, Jerusalem. This find, dating back to the first century, demonstrates that victims of crucifixion could receive proper Jewish burials, supporting the Gospel accounts of Jesus' burial.[51]

5. **The Nazareth Decree**: This is an edict of the Roman Emperor ordering capital punishment for anyone caught disturbing tombs or moving bodies from them. Discovered in Nazareth, this decree may reflect a response to the empty tomb story of Jesus and corroborates the New Testament narrative.[52]

6. **The 'Jesus Boat'**: A well-preserved fishing boat from the time of Jesus was discovered near the Sea of Galilee. This boat aligns with the types of boats described in the Gospels and confirms the fishing culture of the region during Jesus' time.[53]

7. **Pontius Pilate's Inscription**: A plaque uncovered in Caesarea Maritima contains an inscription mentioning "Pontius Pilatus, Prefect of Judea." This discovery provides tangible evidence for the existence of Pontius Pilate, the Roman governor who presided over Jesus' trial.[54]

8. **Evidence for the Crucifixion**: The discovery of the gravesite in Jerusalem containing the remains of individuals who died brutal deaths, including crucifixion, during the Jewish revolt against Rome in 70 A.D., provides physical evidence supporting the Gospel accounts of crucifixion.[55]

9. **Historical Accuracy of Luke**: Archaeological discoveries have affirmed the accuracy of various locations, events, and people described in the Gospel of Luke and the Acts of the Apostles, confirming Luke's credibility as a historian.[56]

---

[51] Zias, J., & Sekeles, E. (1985). *The Crucified Man from Giv'at ha-Mivtar: A Reappraisal.* Israel Exploration Journal, 35(1), 22-27.
[52] Evans, C. A. (2012). *Jesus and His World: The Archaeological Evidence.* Westminster John Knox Press.
[53] Wachsmann, S. (1995). *The Sea of Galilee Boat: An Extraordinary 2000 Year Old Discovery.* New York: Plenum Press.
[54] Ibid 52.
[55] Heschel, S. (2009). *Crucifixion in the Ancient World and the Folly of the Message of the Cross.* Fortress Press.
[56] Hemer, C. J. (1990). *The Book of Acts in the Setting of Hellenistic History.* Edited by Conrad H. Gempf. Tübingen: J.C.B. Mohr (Paul Siebeck).

These discoveries not only validate specific details mentioned in the New Testament but also provide a richer understanding of the historical and cultural context of the time of Jesus.

While these archaeological findings do not prove the theological claims of the Gospels, they do corroborate many of the historical and cultural details described in the narratives. The evidence supports the Gospels as reliable sources of historical information about the time and place in which the events they describe are set. It is important to note that archaeology can only provide a partial picture. Many of the specific events and figures in the Gospels cannot receive direct verification through archaeology. The field helps build a general picture of the historical and cultural background but often cannot confirm specific details.

Archaeological discoveries have played a significant role in corroborating the historical and cultural context of the Gospels. From confirming the existence of critical figures like Pontius Pilate to revealing the living conditions in regions where Jesus lived and ministered, these findings provide a tangible backdrop to the Gospel narratives, enhancing our understanding of the historical setting of these texts.

Comparative Religious and Cultural Studies

Studies of religious and cultural practices of the time have often aligned with the descriptions in the Gospels. The descriptions in the Gospels align with what is known from historical and archaeological sources about first-century Judaism, which contain numerous references to Jewish rituals and customs. One example is the observance of Jewish feasts and festivals, such as Passover and the Feast of Tabernacles. The Gospels not only mention these festivals but also provide details consistent with known Jewish practices of the time. The Gospels accurately depict references to ritual practices like circumcision, purification rites, and Sabbath observance, which are integral to Jewish religious life.

Historical and archaeological findings align with the descriptions of the Temple in Jerusalem as provided by the Gospels. The portrayal of the Temple's role as the center of Jewish worship and the site of major religious festivals is accurate, including details about the Temple's architecture, courts, and the presence of money changers and animal sellers, which aligns with historical understandings of the Temple's structure and function.

The religious diversity of the time is reflected in the Gospels, which describe various Jewish religious groups such as the Pharisees, Sadducees, and Essenes. According to the Gospels, the Pharisees are portrayed as a group that emphasizes strict adherence to the Law and oral traditions, which is consistent with historical accounts of their focus on legalism and interpretations of the Torah. Their portrayal in ways consistent with their historical role and beliefs, particularly their rejection of the resurrection and focus on the Temple, depicts the Sadducees as a priestly, aristocratic group. The religious landscape of the period can be understood by looking at other sources like the Dead Sea Scrolls, which provide information about the presence and practices of the Essenes, although they are not directly

Chapter 4 – Did Jesus Die? – Examining the Historical Evidence

mentioned in the Gospels.

The Gospels reflect Roman-occupied Judea's complex societal and political context, depicting the tensions between Jewish groups and Roman authorities, corroborated by historical sources. The Gospels provide descriptions of Roman taxation, military presence, and governance that are consistent with what we know about Roman provincial administration.

Comparative studies have shown that the linguistic style of the Gospels is consistent with first-century Jewish and Greco-Roman writings. The use of Aramaic and Greek phrases, idiomatic expressions, and literary forms like parables and aphorisms aligns with the linguistic milieu of the region. The Gospels also depict interactions between Jewish communities and neighboring Gentile populations. These interactions reflect the cultural and religious diversity of the region, as well as the influence of Hellenistic culture, which is supported by historical evidence. The portrayal of women and marginalized groups in the Gospels reflects the cultural attitudes of the time. The presence and role of women in Jesus' ministry and the inclusion of tax collectors and sinners align with historical understandings of social hierarchies and norms.

Comparative religious and cultural studies have shown that the Gospels' accounts of Jewish rituals, Temple practices, and various religious groups are consistent with historical knowledge of these practices and groups. This alignment provides a broader context for understanding the Gospels as religious texts and historical documents reflecting the complex religious and cultural landscape of first-century Judea and the Roman Empire.

Early Christian Writings Outside the Gospels

The Apostolic Fathers were early Christian writers who followed the apostles. Their letters and writings in the late 1st and early 2nd centuries provide valuable insights into the early Christian community and its beliefs. Writings like those of Clement of Rome[57], Ignatius of Antioch,[58] and Polycarp of Smyrna[59] often make indirect references to the teachings and events described in the Gospels. For example, Clement's letter to the Corinthians, written around the end of the 1st century, echoes themes and teachings found in the Gospels. In his letters, Ignatius emphasizes themes like the divinity of Christ and the importance of unity in the Church, reflecting beliefs and values central to the Gospel narratives.

Works of early Church historians, such as Eusebius of Caesarea,[60] provide a historical account of the early Christian Church and offer corroborating information about the events and figures mentioned in the

---

[57] Jefford, C. N. (Ed.). (1999). *The Apostolic Fathers: An Essential Guide*. Nashville, TN: Abingdon Press.
[58] Foster, P. (2007). *The Writings of the Apostolic Fathers*. London: T&T Clark.
[59] Hartog, P. (2015). *Polycarp and the New Testament: The Occasion, Rhetoric, Theme, and Unity of the Epistle to the Philippians and its Allusions to New Testament Literature*. Tübingen: Mohr Siebeck.
[60] Eusebius of Caesarea. (1989). *The History of the Church: From Christ to Constantine*. Translated by G. A. Williamson. Revised and updated by Andrew Louth. Penguin Classics.

Gospels. Eusebius, in his "Ecclesiastical History," written in the early 4th century, compiles a history of the early Church, including references to Jesus and the apostles, as well as to events described in the New Testament. Eusebius' work is precious for understanding the development of the early Christian Church and its connection to the apostolic era.

The Didache,[61] also known as "The Teaching of the Twelve Apostles," is an early Christian treatise that provides insights into Christian practices and teachings in the first century. The Didache includes instructions on Christian ethics, rituals like baptism and the Eucharist, and church organization, reflecting the lifestyle and beliefs of early Christians as described in the New Testament. These early Christian writings often corroborate the events, teachings, and figures mentioned in the Gospels, though they do not always provide direct or detailed accounts. They refer to the critical teachings of Jesus, the significance of his death and resurrection, and the work of the apostles, reinforcing the historical framework of the New Testament.

Beyond providing historical corroboration, these writings offer insights into how the early Christian community interpreted and practiced the teachings of Jesus. They show how the early Christian community understood, interpreted, and lived out the teachings of Jesus and the apostles, providing context for developing Christian doctrine and practice. While these writings are invaluable for historical and theological research, they have limitations. These writings were primarily intended for theological instruction, exhortation, or defense of the faith, rather than for historical documentation. The time gap between the events of the Gospels and these writings means they rely on transmitting tradition and teachings.

Early Christian writings outside the Gospels, such as the letters of the Apostolic Fathers and the works of early Church historians, provide additional context and corroboration for the events and figures mentioned in the Gospels. These texts reinforce the historical framework of the New Testament and offer insights into the beliefs, interpretations, and practices of the early Christian community. They serve as a bridge between the apostolic era and the later development of Christian thought and organization.

Documentary and Textual Analysis

Textual scholars scrutinize the writing style of the Gospels to glean insights about their authorship and historical context. This included examining the literary structure, narrative techniques, and rhetorical devices used in the texts. For instance, parables, a standard teaching method in Jewish tradition, are prevalent in the Gospels, particularly in synoptic (Matthew, Mark, and Luke). The narrative style of each Gospel reflects different theological emphases and audience targets, which provides clues about the circumstances and purposes behind their composition.[62]

---

[61] Audet, J. P. (1958). *La Didachè: Instructions des Apôtres.* Paris: J. Gabalda.
[62] Burridge, R. A. (2004). *What Are the Gospels? A Comparison with Graeco-Roman Biography.* 2nd ed. Grand Rapids, MI: Eerdmans.

Linguistic analysis involves examining the language used in the Gospels and comparing it to other contemporary texts. This reveals essential aspects such as the predominant use of Koine Greek in the Gospels, a standard dialect in the Hellenistic world, which suggests a target audience beyond just the Jewish community. Aramaic terms and phrases in the Gospels indicate a connection to the language spoken in Judea during the 1st century, adding to their historical authenticity.[63]

Scholars pay close attention to specific terms and phrases used in the Gospels, which shed light on the historical and cultural context of the narratives. For example, titles used for Jesus, such as "Rabbi" or "Son of Man," reflect Jewish cultural and religious contexts.[64] References to specific social customs, political structures, and geographical locations help to authenticate the texts' historical setting. Comparing the Gospels with other writings from the same period, both within and outside the Christian tradition, helps assess their historical plausibility. This includes analyzing similarities and differences in themes, content, and style with other Jewish and Hellenistic writings and comparing the Gospels to non-biblical accounts of the period to corroborate or contrast the historical details mentioned.[65]

Examining textual variants among different Gospel manuscripts helps understand the transmission and preservation of these texts. Scholars analyze variations in wording, omissions, or additions across different manuscript traditions, the age and origin of manuscripts to trace the development of the Gospel texts over time.[66] Scholars play a crucial role in determining the approximate dates when the Gospels were written during textual analysis. Factors considered include references to historical events (e.g., the destruction of the Jerusalem Temple in 70 CE) or the development of early Christian theology and ecclesiology as reflected in the texts.[67]

By understanding the socio-political context in which the Gospels were written, we can gain insights into the emphasis placed on specific themes or narratives.[68] This includes the impact of Roman occupation on Jewish society and the early Christian community and the relationship between the emerging Christian movement and mainstream Jewish practices and beliefs. Documentary and textual analysis of the Gospels is a multifaceted approach that examines the writing style, language, specific terms, and the historical context of these texts. It compares the Gospels with other contemporary writings to assess their historical authenticity and understand their place within the time's broader literary and cultural landscape. This analysis is crucial in piecing together the Gospels' historical reliability and

---

[63] Porter, S. E. (1997). *The Language of the New Testament: Classic Essays*. Sheffield: Sheffield Academic Press.
[64] Evans, C. A. (2012). *Jesus and His World: The Archaeological Evidence*. Westminster John Knox Press.
[65] Meier, J. P. (1991). *A Marginal Jew: Rethinking the Historical Jesus, Volume I: The Roots of the Problem and the Person*. New York: Doubleday.
[66] Ibid 14.
[67] Robinson, J. A. T. (1976). *Redating the New Testament*. Philadelphia: Westminster Press.
[68] Horsley, R. A. (2001). *Bandits, Prophets, and Messiahs: Popular Movements at the Time of Jesus*. Harrisburg, PA: Trinity Press International.

understanding the development of early Christian thought and tradition.

Non-Christian Religious Texts

Some non-Christian texts from the era, including Jewish and Gnostic writings, provide a broader context for understanding the religious and philosophical environment in which Christianity emerged. These texts at times allude indirectly to figures or movements that have connections to the early Christian community. Jewish texts from the Second Temple period, such as the Dead Sea Scrolls,[69] provide valuable context for understanding the religious environment in which Christianity emerged. These writings illuminate various Jewish sects, beliefs, and practices of the time, some of which the New Testament reflects, including interpretations of Hebrew Scriptures that can be compared with those found in the Gospels, offering insights into different Jewish theological perspectives of the era.[70]

Later rabbinic texts, like the Mishnah and Talmuds, while compiled after the New Testament period, contain traditions and teachings that go back to the time of Jesus. These texts offer a glimpse into Jewish religious thought and legal interpretations that were contemporary with the early Christian period.[71] These texts serve as a contrast to many teachings of Jesus in the Gospels, providing a background on Pharisaic and rabbinic Judaism.[72]

Gnostic writings, such as those in the Nag Hammadi library,[73] represent a different religious and philosophical stream from early Christianity. These texts highlight the diversity of thought in the early Christian era, particularly in terms of cosmology, soteriology, and Christology, offering a different perspective on Jesus and his teachings, which we can contrast with the orthodox portrayals in the Gospels.[74]

The writings from the broader Greco-Roman world in which Christianity took root provide context for the cultural and philosophical milieu of the era.[75] These include works by philosophers, historians, and playwrights. These texts help understand the prevailing philosophical and religious ideas that early Christianity interacted with and sometimes incorporated or rebutted. They provide a backdrop for the spread of Christianity in the Roman Empire and the subsequent cultural and religious interactions.[76]

Some of these non-Christian texts indirectly reference early Christian

---

[69] Vermes, G. (1997). *The Complete Dead Sea Scrolls in English*. New York: Penguin Books.
[70] Charlesworth, J. H. (Ed.). (2010). *The Old Testament Pseudepigrapha and the New Testament: Prolegomena for the Study of Christian Origins*. Cambridge: Cambridge University Press.
[71] Neusner, J. (1988). *Judaism and Christianity in the Age of Constantine: History, Messiah, Israel, and the Initial Confrontation*. Chicago: University of Chicago Press.
[72] Cohen, S. J. D. (1987). *From the Maccabees to the Mishnah*. Philadelphia: Westminster Press.
[73] Robinson, J. M. (Ed.). (1990). *The Nag Hammadi Library in English*. 3rd, Completely Revised Edition. San Francisco: HarperSanFrancisco.
[74] Pagels, E. (1979). *The Gnostic Gospels*. New York: Random House.
[75] Hengel, M. (1989). *Judaism and Hellenism: Studies in Their Encounter in Palestine During the Early Hellenistic Period*. 2 Volumes. Philadelphia: Fortress Press.
[76] MacMullen, R. (1984). *Christianizing the Roman Empire (A.D. 100-400)*. New Haven: Yale University Press.

Chapter 4 – Did Jesus Die? – Examining the Historical Evidence

figures or movements. For example, the Jewish texts refer to the early followers of Jesus in a manner that reflects the Jewish perspective on this emerging sect. On the other hand, Greco-Roman authors occasionally mention Christians, though often in a context of misunderstanding or persecution. By comparing the New Testament texts with these non-Christian writings, scholars help gain a more nuanced understanding of the religious and philosophical diversity of the time and a better understanding of how Christianity differentiated itself from contemporary Jewish and pagan beliefs.

We need to realize that non-Christian religious texts from the era surrounding the emergence of Christianity provide a broader context for understanding the religious and philosophical environment of the time. These texts, from Jewish writings to Gnostic and Greco-Roman works, offer insights into the diverse beliefs and practices that coexisted with early Christianity. They help to illuminate the historical backdrop against which Christianity developed and differentiated itself from other contemporary religious and philosophical movements.

If we were to create a more organized list of non-Christian texts that provide evidence of Jesus' death, we would need to include some of the texts we've already discussed, as well as other relevant documents.

1. **Tacitus' Annals**: Roman historian Tacitus, in his work "Annals," written around 116 AD, refers to Jesus' execution. He mentions Jesus was executed during the reign of Emperor Tiberius, under the authority of Pontius Pilate. This is one of the earliest non-Christian references to Jesus and is significant for its independent corroboration of the crucifixion.[77]

2. **Flavius Josephus' Antiquities of the Jews**: Jewish historian Josephus, in "Antiquities of the Jews," includes a passage known as the Testimonium Flavianum. This passage mentions Jesus, his reputation as a wise and virtuous man, his crucifixion at the hands of Pontius Pilate, and the continued following of his disciples. Most scholars agree that the entire passage's authenticity is debated, but they do agree that it originally referenced Jesus' execution.[78]

3. **The Babylonian Talmud**: The Talmud, a central text of Rabbinic Judaism, includes references to Yeshu (a name associated with Jesus in Jewish texts) and his execution. The Talmudic references are indirect and subject to interpretation, but they are generally understood as alluding to Jesus and his death. However, these texts were compiled much later than Jesus' time by people who added more legendary elements.

---

[77] Ibid 47.
[78] Ibid 48.

4. **Lucian of Samosata**: Lucian, a Greek satirist, wrote about Christians in the mid-2nd century and referred to Jesus indirectly. He mentioned that Christians worshiped a man who brought new teachings and was crucified in Palestine. Despite its satirical and critical nature towards Christianity, Lucian's account confirms that Jesus was recognized as a historical figure who was crucified.[79]

5. **Mara Bar-Serapion's Letter**: a Syrian philosopher, wrote a letter to his son, possibly sometime after 73 AD, making a reference to the execution of the "wise king" of the Jews by his own people. While not explicitly naming Jesus, many scholars believe this is a likely reference to him, considering the context and timing of the letter.[80]

6. **Pliny the Younger's Letters**: Pliny the Younger, a Roman governor, wrote to Emperor Trajan seeking advice on how to deal with Christians. Pliny the Younger's letters, dated around 112 AD, imply knowledge of Jesus as the central figure of Christianity, around whom the early Church rituals and beliefs were centered, even though they don't directly mention Jesus' death.[81]

7. **Thallus and Phlegon**: Later Christian writers cite Thallus and Phlegon, both first-century historians, as mentioning events that could be linked to the darkness and earthquake described in the Gospels at the time of the crucifixion. However, their original works got lost, and later authors only know about the references through citations.[82][83]

8. **The Gnostic Gospels**: While not historical texts, the Gnostic Gospels, such as the Gospel of Thomas, offer insight into early Christian thought and sometimes include references to Jesus' death. These texts, however, are more theological and allegorical rather than historical accounts.[84]

One more source that we mentioned earlier refers to the Talmud writings. The compilation of the Talmud took place several centuries after the time of Jesus, and it reflects the rabbinic Jewish perspective of that later period. The texts mentioning Jesus are often cryptic, indirect, and subject to varying interpretations. Here are some notable passages that some scholars have interpreted as referring to Jesus and his death:

---

[79] *Lucian of Samosata*. (Mid-2nd century). The Death of Peregrine.
[80] Bar-Serapion, M. (73 AD or later). *Mara Bar-Serapion's Letter*.
[81] Pliny the Younger. (112 AD). *Letters (Vol. 1)*. (B. Radice, Trans.). Harvard University Press.
[82] Carrier, Richard (2011–2012). "Thallus and the Darkness at Christ's Death" (PDF). The Journal of Greco-Roman Christianity and Judaism. 8: 185–191.
[83] Roberts, Donaldson & Coxe (1896), Volume IV, *"Contra Celsum"*, Book II, chapter 14,23,59 p. 441.
[84] Meyer, Marvin. *The Nag Hammadi Scriptures*: The International Edition. HarperOne, 2007. pp. 2–3. ISBN 0-06-052378-6

Chapter 4 – Did Jesus Die? – Examining the Historical Evidence

1. **Sanhedrin 43a**: This passage is one of the most frequently cited in discussions about Jesus in the Talmud. It describes the execution of "Yeshu," a common Talmudic name for Jesus. The text mentions that Yeshu was hanged on the eve of Passover, which aligns with the Gospel account of Jesus' crucifixion occurring around this time. The passage also states that a herald called out for forty days seeking witnesses in defense of Yeshu before his execution, suggesting a trial or an opportunity for rebuttal.

2. **Sanhedrin 107b and Sotah 47a**: mention a certain "Yeshu" who they describe as a student of Rabbi Joshua ben Perachiah and later being led astray into idolatrous practices. The identification of this Yeshu with Jesus of Nazareth sparks debate among scholars, who argue that the context and period described must align with the New Testament accounts.

3. **Gittin 56b-57a**: This passage describes the destruction of Jerusalem and includes a story about "Balaam," which is sometimes interpreted as a derogatory code for Jesus in the Talmud. Within the Talmud, the text is part of the broader discourse where his story is interpreted negatively, although it does not directly mention the death of Jesus.

4. **Various References to Sorcery and Heresy**: In several instances, the Talmud refers to a figure named Yeshu or Balaam in the context of sorcery, misleading the people or heresy. Scholars often see these references as controversial responses to the growing influence of Christianity, reflecting the rabbinic perspective of the time.

Conclusion

External evidence that comes from a variety of sources supports the Gospel narratives. These include historical writings, archaeological discoveries, comparative studies, and other early Christian texts. Although every detail of the Gospels is not validated by this evidence, it does offer a framework that supports many aspects of the historical and cultural context in which the events of the Gospels are said to have taken place. This external evidence is essential for understanding the Gospels as religious texts and historical documents within their broader 1st-century context. By analyzing this external evidence, scholars and researchers can gain a more comprehensive understanding of the Gospels' origins, themes, and significance. Therefore, it is crucial to consider external evidence when studying and interpreting the Gospels.

Chapter 5

Did Jesus Die? – Examining the Medical Evidence

Physical Endurance of Jesus

To demonstrate the possibility of Jesus' death, we need to examine his overall health and physical abilities prior to the day of his crucifixion. However, it is crucial to avoid assuming that Jesus could never have died because he was divine. Instead, we must analyze and consider the writings of the Gospels to assess Jesus' lifestyle and physical endurance.

Since we cannot access Jesus' medical records, our analysis is limited to the accounts of his life and ministry as described in the Gospels. In the previous chapter, we discussed the reliability and historicity of the Gospels as evidence for Jesus' life and death. After conducting much analysis, most critics have recognized the Gospels as a reliable historical account.

In order to determine Jesus' physical endurance, we will analyze the accounts of his extensive travels for his ministry, often on foot, across various regions such as Galilee, Judea, and Jerusalem. This aspect of his life highlights several key points that are relevant to our investigation. For instance, it suggests that Jesus was physically fit and capable of enduring long journeys on foot, which would have required significant stamina and resilience. Additionally, it implies that Jesus was likely exposed to a variety of environmental and physical stressors during his travels, such as extreme temperatures, rugged terrain, and potential threats from hostile groups, which could have further tested his physical abilities and resilience. By examining these aspects of Jesus' life and ministry, we can gain valuable insights into his overall physical health and endurance, which are essential to understanding the plausibility of his death by crucifixion.

Extensive Traveling: The travels of Jesus, as described in the New Testament, encompassed a diverse array of geographical locations, each

Chapter 5 – Did Jesus Die? – Examining the Medical Evidence

presenting its challenges and requiring notable physical endurance. Jesus' journeys took him through varied landscapes, from the hilly and mountainous regions of Galilee to the more arid areas around Jerusalem and Judea. Navigating these terrains, especially on foot, would have demanded high physical fitness and adaptability.

The distances between the cities, villages, and rural areas he visited were significant, particularly in an era without modern transportation. He would often travel these significant distances between cities, villages, and rural areas on foot, occasionally using a boat to cross the Sea of Galilee, showcasing his physical stamina and unwavering commitment to his mission. Traveling in the ancient Middle East involved exposure to various environmental conditions – hot, dry climates, dust storms, and variable weather conditions. Coping with these elements would have required robust physical health and resilience.

The travels of Jesus and his companions are a fascinating subject to study, as they covered a wide range of distances and terrains across the region. To better understand the journeys of Jesus, we need to look at the estimated distances between various locations that he visited. While these distances are not exact, they provide a general idea of the extent of his travels.

According to historical records, Jesus traveled from Nazareth to Bethlehem, which is estimated to be a distance of 90-100 miles (145-160 kilometers). This journey likely involved traveling through the hills of the Lower Galilee region before descending to the Sea of Galilee. Jesus also traveled from Nazareth to Capernaum, which is estimated to be a distance of 20-25 miles (32-40 kilometers) through the hilly terrain of the Lower Galilee region.

From Capernaum, Jesus traveled to Jerusalem, covering an estimated distance of 80-85 miles (130-137 kilometers). This journey involved passing through the Jordan River Valley, and depending on the route chosen, passing through areas like Samaria or the Eastern Jordan Valley before ascending to the Judaean hills to reach Jerusalem.

Once in Jerusalem, Jesus traveled a short distance of 2 miles (3.2 kilometers) to reach Bethany, which was a quick walk on the eastern slope of the Mount of Olives. From Jerusalem, Jesus also traveled to Jericho, which is estimated to be a distance of 18-20 miles (29-32 kilometers). This journey involved a descent from the highlands of Jerusalem to the below-sea-level city of Jericho, located in the Jordan River Valley.

Jesus and his disciples also traveled around the Sea of Galilee, which is approximately 13 miles (21 kilometers) long and 8 miles (13 kilometers) wide. They frequently traveled by boat across the sea to reach various towns and villages along its shores.[1]

These distances highlight the extent of Jesus' travels across various terrains, mostly on foot, connecting him with diverse populations and allowing his teachings to reach people in different social and economic

---

[1] Rasmussen, C. G. (2013). *Zondervan Essential Atlas of the Bible*. Zondervan.

contexts. By studying the estimated distances of Jesus' travels, we can gain a better understanding of the geographical and historical context of his teachings and the impact they had on the region.

During these travels, the explorers would have faced limited and variable quality resources like food, water, and shelter. Maintaining physical strength and health in the face of such limitations indicates considerable physical robustness. In addition to the physical act of traveling, Jesus was constantly engaging with people, teaching, and often dealing with large crowds. This constant interaction demanded physical energy and mental and emotional endurance.

Traveling across different regions also meant encountering various cultural and social contexts, each with its norms and expectations. Navigating these effectively would require physical stamina and a high level of social and emotional intelligence. The extensive travels of Jesus described in the New Testament underscore a figure of remarkable physical and mental endurance. The ability to traverse long distances on foot across challenging terrains, coupled with the demands of his ministry, paints a picture of a highly resilient individual in both physical and psychological aspects.

Teaching and Preaching: Besides his extensive travels, the New Testament depicts Jesus as actively engaged in teaching and preaching, which would have demanded a significant amount of physical, mental, and emotional energy. Jesus often spoke to large gatherings ranging from small groups to thousands of people. Addressing such diverse audiences, sometimes in open and challenging environments like hillsides or near lakes, would require a strong voice and the stamina to speak for extended periods.

He taught in various types of venues, without confining himself to any single one. He taught in synagogues, private homes, outdoor spaces, and even from boats. Each setting presented challenges, from managing acoustics to ensuring that he effectively communicated his message to every audience member. The content of Jesus' teachings, which included parables, discussions of religious law, ethical teachings, and debates with religious leaders, demanded a high degree of intellectual and rhetorical skill. Articulating complex theological and ethical concepts in ways that were accessible and compelling to diverse audiences would require significant mental acuity.

Jesus' teachings often touched on profoundly personal and emotional subjects, including morality, faith, forgiveness, and compassion. Engaging with people on such a level, often responding to their questions or doubts, would involve considerable emotional intelligence and empathy. Alongside his teachings, Jesus counsels individuals and performs healings, adding to the emotional and physical demands of his ministry. These activities would require physical touch a d, a deep sense of compassion, and the ability to connect with people in their moments of vulnerability.

Jesus faced skepticism and direct opposition from various groups, including religious authorities and skeptics among the public. Handling such confrontations, maintaining his composure, and effectively communicating his message in the face of opposition would require mental resilience and a calm demeanor. Jesus managed to consistently maintain his teaching and preaching schedule despite the challenges. This consistency highlights physical stamina and a strong, unwavering commitment to his mission and message.

As described in the New Testament, the teaching and preaching aspects of Jesus' life illustrate a figure of remarkable endurance and versatility. These activities required the physical ability to travel and speak to large crowds, the mental sharpness to engage in theological discussions, and the emotional capacity to connect with and address the needs of diverse individuals.

Rugged Terrain and Long Journeys: The New Testament narratives describe Jesus' traversing regions like Galilee and Judea, known for their challenging landscapes and a considerable distance between locations. This aspect of his life underscores several vital points. Galilee, characterized by its hills and valleys, and Judea, with its more arid and rocky terrain, presented different physical challenges. Moving through these areas, especially on foot, would have required adaptability to varied topographical features.[2]

The distances between towns and villages in these regions were not insignificant, particularly in an era devoid of modern transportation. Journeying these distances on foot, a standard mode of travel at the time, would necessitate considerable physical endurance and resilience. Traveling in the open landscapes of these regions meant exposure to the elements – harsh sun, wind, and possibly inclement weather. Such conditions could be physically taxing and demand a solid constitution to withstand potential dehydration, heat exhaustion, or cold.

The paths Jesus and his followers would have taken were likely unpaved and uneven, traversing through hills and valleys and possibly crossing rivers or streams. Such journeys would be physically demanding, requiring good balance, strength, and stamina. As was typical for itinerant teachers and their followers in those times, traveling with minimal resources meant limited access to food, water, and shelter. Managing these scarcities effectively would indicate high physical robustness and resourcefulness.

The journeys across different regions also involved interactions with a wide array of people from various social, cultural, and economic backgrounds. This required physical, social, and emotional energy to engage effectively with different communities. While the New Testament focuses on the active aspects of Jesus' travels, it's reasonable to assume that rest and recovery periods were essential, particularly after traversing rugged terrains and long distances. Managing physical exertion with adequate rest would be crucial for maintaining overall health.

---

[2] *The Physical Geography, Geology, and Meteorology of the Holyland* by Henry Baker Tristram 2007 ISBN 1593334826 page 11

The travels of Jesus across the rugged terrains and long distances of Galilee and Judea highlight a figure of significant physical strength and endurance. The ability to undertake such demanding journeys, often with limited resources and under challenging environmental conditions, underscores a level of physical fitness and adaptability that was integral to his itinerant ministry.

Lifestyle Indications: The Gospels present a picture of Jesus leading a modest lifestyle characterized by minimal material possessions and focusing on spiritual and mission-driven objectives. This lifestyle likely played a significant role in supporting his physical endurance and overall health. Jesus is portrayed as living a simple life and not owning many personal belongings. This minimalist approach meant fewer physical burdens, allowing for more accessible travel and less distraction from his mission.

The Gospels often mention Jesus relying on the hospitality of others for food and shelter. This way of living, while humble, ensured that he could travel without carrying extensive provisions, thus reducing physical strain. While specific details are scarce, it's plausible that Jesus' diet was typical of the region and time – largely plant-based with occasional fish and meat, grains, fruits, and vegetables. Such a balanced and natural diet would be conducive to maintaining good health and energy levels.

His day-to-day activities, including walking long distances and engaging in manual labor (as he is often referred to as a carpenter), would have contributed to Jesus' physically active lifestyle. Regular physical activity is known to improve overall health and stamina. The intense mental and spiritual focus of Jesus' life, as described in the Gospels, might have also contributed positively to his physical health. Mind-body connections are well-documented in health studies, and a sense of purpose and spiritual fulfillment can benefit physical well-being.

Despite the significant stress he faced, especially towards the end of his life, Jesus is often depicted as seeking solitude for prayer and reflection. These practices could have been crucial for managing stress, allowing for mental and emotional recovery, which is essential for physical health. The presence of disciples and followers provided a support network, offering companionship and practical support during travels and ministries. This sense of community could have been a source of emotional strength and resilience.

The lifestyle of Jesus, as portrayed in the Gospels, suggests a balance of physical activity, modest living, mental and spiritual focus, and community support. This combination likely played a significant role in sustaining his physical endurance and health, enabling him to carry out his extensive travels and ministry effectively.

Overall Health: The depiction of Jesus in the Gospels, particularly his capacity for extensive travel and continuous teaching, suggests that he maintained a robust level of overall health for the majority of his ministry. There are several aspects in which we can expand upon this observation.

Chapter 5 – Did Jesus Die? – Examining the Medical Evidence

The physical demands of frequent traveling, often on foot, over long distances, and through challenging terrains, as well as the rigors of public speaking and teaching, indicate a high level of physical endurance and stamina. Such abilities are typically associated with good cardiovascular and muscular health.

There are no accounts in the Gospels suggesting that Jesus suffered from any chronic illnesses. While the absence of such information is not definitive proof, it contributes to his portrayal as a physically capable individual. The mental and emotional demands of Jesus' role – including dealing with crowds, confronting skeptics and opponents, and providing counsel and healing – would require physical and psychological resilience. His ability to handle these stresses implies a solid mental and emotional constitution.

Jesus' lifestyle involved adapting to different environments, weather conditions, and varying food and water availability levels. His ability to thrive in such varying conditions suggests a robust physical state. Jesus actively participates in spiritual practices like prayer and meditation, which can have a positive impact on overall health. These practices contributed to stress reduction and mental well-being, complementing his physical health.

It's important to note that the Gospels detail a significant decline in Jesus' physical state leading up to and during the crucifixion. This event, marked by extreme physical and emotional trauma, stands in contrast to the otherwise robust health he seemed to maintain. We must acknowledge that the Gospels, which serve as the source of these observations, were written with spiritual and theological intentions rather than as historical or medical documentation. Therefore, the descriptions of Jesus' health and physical capabilities may also serve symbolic or illustrative purposes in the context of his teachings and the Christian faith.

The overall health of Jesus, as inferred from the Gospel narratives, appears to have been quite good, supporting his intensive ministry activities. However, it is important to interpret these descriptions within their religious and theological context and acknowledge that they may not have been intended as literal, historical records of his physical condition.

The chronologic events on the day of Jesus crucifixion

In order to gain a deeper understanding of Jesus' physical condition leading up to his crucifixion, as described in the Gospels, we can piece together a chronological sequence of events from Good Friday. This timeline, derived from the accounts in the four Gospels, helps to illustrate the physical and emotional toll these events likely had on Jesus, providing insight into his overall medical condition and the factors that contributed to his rapid deterioration and subsequent death.

1. **Night to Early Morning**:
    - **Last Supper**: Jesus has the Last Supper with his disciples, where he predicts his betrayal.

- **Gethsemane**: Jesus goes to the Garden of Gethsemane to pray. He undergoes agony and Judas Iscariot betrays him.
- **Arrest**: temple guards arrest Jesus.

This is the beginning of physical and psychological exertion. After the Last Supper, Jesus goes to the Garden of Gethsemane on the Mount of Olives near Jerusalem. His disciples accompanied him, indicating the garden was a familiar place for prayer and reflection. In the garden, Jesus experiences profound emotional distress. The Gospels, particularly Luke's account, describe him as being in agony. He prays fervently, expressing deep sorrow and asking if it is possible for the impending suffering to be avoided, yet submitting to God's will. This moment highlights Jesus' humanity and the weight he is about to endure.

Jesus asks Peter, James, and John to stay awake and pray. However, they fall asleep, adding to his sense of isolation and burden. While Jesus is still in the garden, Judas Iscariot arrives with a crowd armed with swords and clubs sent by the chief priests and elders. Judas betrays Jesus with a kiss, a prearranged sign to identify Jesus to the authorities. Jesus responds to the betrayal with calm and dignity, addressing Judas and then speaking to the crowd, acknowledging that his arrest is unfolding as foretold. He heals the ear of a servant of the high priest, which was cut off during a brief scuffle initiated by one of his disciples. After this, the authorities took Jesus into custody, signaling the start of a series of trials that would eventually lead to his crucifixion.

2. **Early Morning Trials**:
    - **Before Ananias and Caiaphas**: the authorities take Jesus to Ananias, a former high priest, and then to Caiaphas, the current high priest, for interrogation and condemnation.
    - **Peter's Denial**: During this time, Peter denies knowing Jesus three times.
    - **Before the Sanhedrin**: the Jewish ruling council tried Jesus and found him guilty of blasphemy.

After Ananias interrogates, Jesus is taken to Caiaphas, the high priest, where he is brought before the Sanhedrin, the Jewish ruling council. This setting is more formal and is essentially a religious trial to determine Jesus' adherence to Jewish law. Caiaphas and the members of the Sanhedrin question Jesus, primarily focusing on his identity and claims of being the Messiah, the Son of God. This interrogation is intense, as the high priest seeks to find grounds for blasphemy, a serious charge under Jewish law.

They bring forward false witnesses to testify against Jesus, but their testimonies do not align. The aim is to find a legitimate accusation to use against Jesus, but inconsistencies and legal irregularities mark the process. Initially, Jesus remains silent in the face of allegations, which fulfills prophecies about the Messiah's suffering. His response, or lack thereof, further frustrates the high priest and the council.

Chapter 5 – Did Jesus Die? – Examining the Medical Evidence

The turning point comes when Caiaphas asks Jesus if he is the Son of God. Jesus affirms this, which the Sanhedrin deems blasphemous. This affirmation leads to his formal condemnation by the Jewish authorities. Following his condemnation, Jesus suffers physical abuse at the hands of those present. The Gospels describe Jesus is mocked, struck, and spat upon by some of the temple guards and others present (Matthew 26:67). This abuse likely included slapping or hitting with fists and possibly the use of objects to strike him. Such treatment would have been both physically painful and deeply humiliating.

Alongside the physical abuse, individuals mock Jesus about his prophetic abilities, further adding to the humiliation. This ridicule is particularly poignant given the grave nature of his situation. The physical and emotional abuse that Jesus endured during this trial before Caiaphas was a significant part of his suffering. It highlights not only the physical pain inflicted but also the psychological torment of being rejected, condemned, and mocked by religious authorities and others who had turned against him. This event is a critical component of the Passion narrative, underscoring the depth of Jesus' suffering leading up to his crucifixion.

3. **Early Morning to Midmorning**:
   - **Before Pilate**: Jesus is taken to Pontius Pilate, the Roman governor. Pilate finds no basis for a charge against Jesus.
   - **Before Herod Antipas**: Pilate sends Jesus to Herod Antipas, who is in Jerusalem at the time. Herod, after mocking Jesus, sends him back to Pilate.
   - **Sentenced by Pilate**: Despite Pilate's initial reluctance, the crowd demands Jesus' crucifixion. Pilate finally gives in and sentences Jesus to death.

Initially, the Jewish leaders bring Jesus before Pontius Pilate. They accuse him of various offenses, including claiming to be the king of the Jews, which can be seen as a challenge to Roman authority. Pilate initially appears reluctant to convict Jesus, not finding a basis for the charges against him. According to the Gospels, Pilate does not see Jesus threatening Roman rule and attempts to release him, seeing the issue as an internal matter among the Jews.

Pilate offers the crowd a choice between releasing Jesus or Barabbas, a known criminal. Influenced by the chief priests, the crowd chooses to release Barabbas and insists on crucifying Jesus. Before sending Jesus to be crucified, Pilate orders him to be scourged; a common Roman practice before crucifixion involves whipping the condemned with a flagrum, a whip with multiple leather thongs, often embedded with metal, bone, or glass. This brutal punishment caused deep, lacerating wounds, significant blood loss, and immense pain.

After the scourging, Roman soldiers mock Jesus for his alleged claim to kingship. They clothe him in a purple robe, place a crown of thorns on his head, and give him a reed as a mock scepter. They kneel and hail him in derision as "King of the Jews," striking him on the head with the reed and spitting on him. This phase of abuse is not only physically excruciating but also psychologically torturous. The mocking actions of the soldiers are designed to humiliate Jesus and ridicule his supposed claims. Despite his initial reluctance and his wife's warning about Jesus (as mentioned in Matthew's Gospel), Pilate eventually yields to the crowd's demands. After symbolically washing his hands to show that he is not responsible for Jesus' death, Pilate hands Jesus over to be crucified. The physical abuse that Jesus endured before Pilate is a significant element of his suffering. The scourging and mocking, according to Christian theology, are critical aspects of Jesus' Passion that exemplify the depth of suffering he underwent before the crucifixion.

4. **Late Morning**:
   - **Scourging and Mockery**: Roman soldiers scourge, mock, and crown Jesus with thorns.
   - **Carrying the Cross**: Jesus carries his cross through the streets of Jerusalem towards Golgotha (also called Calvary).

After Roman soldiers scourge, mock, and crown Jesus with thorns, they force Him to carry His cross through the streets of Jerusalem towards Golgotha on Via Dolorosa. The Via Dolorosa, located in the Old City of Jerusalem, traverses a terrain characteristic of a densely built, historical urban setting. The route of the Via Dolorosa is through narrow and winding streets, typical of ancient cities. These streets can be relatively confined and sometimes ascend or descend slightly, given the hilly topography of Jerusalem. Most of the path consists of stone slabs that they paved. These stones can be uneven and, in places, slippery, especially if worn smooth by the passage of countless pilgrims over the centuries.

There are sections with steps and gentle slopes. The hilly terrain of the Old City of Jerusalem serves as the foundation, and although the Via Dolorosa doesn't include steep inclines, it does mirror the undulating nature of the city. The Via Dolorosa runs through an active part of the Old City and is often crowded with pilgrims, tourists, and local residents and traders. This aspect would give the walk a very vibrant and lively atmosphere. Today, the route passes through areas that reflect Jerusalem's rich cultural and religious diversity, with Christian, Jewish, and Muslim quarters in proximity.[3]

According to belief, Jesus walked the Via Dolorosa, a traditional route in Jerusalem, while carrying the cross to his crucifixion. This path meanders for about 0.37 miles (600 meters) through the Old City of Jerusalem, starting from the former location of the Antonia Fortress where Pontius Pilate sentenced Jesus to death, and ending at the Church of the Holy Sepulchre. It is widely believed that this church houses both the locations of Jesus' crucifixion at Golgotha and his tomb.

As with any other condemned prisoner, they forced Jesus to carry the

---

[3] Montefiore, S. S. (2011). *Jerusalem: The Biography*. Alfred A. Knopf.

Chapter 5 – Did Jesus Die? – Examining the Medical Evidence

cross through the slopes of Dolorosa. The weight of the cross that Jesus carried to his crucifixion needs to be specified in the Christian Biblical texts. As a result, its exact weight is a matter of speculation and historical estimation. However, we can derive a reasonable estimate based on historical and archaeological understanding of Roman crucifixion practices and the construction of crosses during that period.

Roman crosses used for crucifixion came in different shapes, such as the traditional Latin cross (crux immissa), the T-shaped cross (crux commissa), and the simple stake (crux simplex). The most commonly depicted form in Christian iconography is the Latin cross. They typically used wood to make crosses, opting for locally available and sturdy materials that could support a human body. This might have included olive lumber, cedar, or cypress in the Jerusalem region.

The cross generally consists of the upright post (stipes) and the crossbeam (patibulum). They would typically permanently fix the upright post in the execution site, while they would force the condemned to carry the crossbeam to the site. Estimates for the weight of the crossbeam suggest it could have been around 75 to 125 pounds (approximately 34 to 57 kilograms). Someone who had endured scourging and was weakened, as Jesus described in the Gospels, would have found this weight to be a significant burden.

If we consider the entire cross, including both the stipes and patibulum, the weight would be considerably more, but it is less likely that the condemned would have carried the whole structure. It's essential to recognize that these estimates are based on historical knowledge and archaeological findings related to Roman execution methods, and no definitive historical source states the exact weight of the cross that Jesus carried. The emphasis in the Biblical accounts is on carrying the cross and its meaning rather than on the specific physical dimensions of the cross.[4]

The exact weather conditions during the time of Jesus' crucifixion remains unknown, as the biblical texts do not provide any specific details about the climate on that day. However, we can make an educated guess about the general climate conditions in Jerusalem around the event, which traditionally observed as Good Friday in the Christian calendar.

It is commonly believed that the crucifixion of Jesus occurred during the Jewish festival of Passover, which falls in late March or April. During this time of year, Jerusalem's climate usually transitions from the rainy winter to the dry, hot summer. The temperature is generally mild to warm, ranging from the high teens to mid-twenties degrees Celsius (around 60-75 degrees Fahrenheit) during the day. Although spring is drier than winter, there is still a possibility of rain. April, in particular, can still have some rainy days, although they are less frequent than in winter.[5]

---

[4] Biblical Archaeology Society. (n.d.). *Roman Crucifixion Methods Reveal the History of Crucifixion.* Retrieved from https://www.biblicalarchaeology.org/daily/biblical-topics/crucifixion/roman-crucifixion-methods-reveal-the-history-of-crucifixion/
[5] Ben-Yoseph, J. (1985). *THE CLIMATE IN ERETZ ISAEL DURING BIBLICAL TIMES.* Hebrew Studies, 26(2), 225–239. http://www.jstor.org/stable/27908940

The Gospels mention specific atmospheric conditions at the time of Jesus' death. For example, the Gospel of Matthew (27:45), Mark (15:33), and Luke (23:44-45) describe darkness falling over the land for several hours in the afternoon. Scholars and theologians has interpreted this in various ways. Some suggest that it could have been a natural phenomenon such as a dust storm, an eclipse (although this is astronomically unlikely during Passover, which coincides with a full moon), or a symbolic or miraculous event.

In conclusion, the exact weather conditions on the day of Jesus' crucifixion are unknown. However, based on the typical climate of Jerusalem during late March or April, we can make an informed guess about the general climate conditions around that time.

According to Christian tradition and various religious texts, people know the path that Jesus took while carrying the cross to his crucifixion as the "Via Dolorosa." The route encompasses fourteen stations, each of which represents an event that occurred along the way. However, it is important to note that the canonical Gospels of the New Testament do not definitively record the number of times Jesus stopped on the Via Dolorosa.

The tradition of Jesus stopping at each station along the Via Dolorosa mainly comes from Christian devotional practices that developed over time. These practices took the form of the Stations of the Cross, also known as the Way of the Cross or Via Crucis.

The Stations of the Cross are a series of images or sculptures representing events from Jesus' journey to his crucifixion. They depict Jesus' condemnation, his carrying of the cross, his falls, his meeting with his mother, his crucifixion, and his burial. The purpose of this devotion is to help followers of Christianity meditate on the Passion of Christ and to connect with the suffering of Jesus.

A detailed description of the route and the stations is available for those who are interested in learning more about this important aspect of Christian tradition:

- **Station I: Jesus is Condemned to Death**
  Location: Near the former site of the Antonia Fortress, where Jesus was tried and condemned by Pontius Pilate.

- **Station II: Jesus Carries His Cross**
  Location: Close to the location of the former Antonia Fortress, where Jesus was given the cross.

- **Station III: Jesus Falls the First Time**
  This station marks the first fall of Jesus under the weight of the cross.

- **Station IV: Jesus Meets His Mother, Mary**
  This station commemorates the encounter between Jesus and his mother Mary on the way to Golgotha.

Chapter 5 – Did Jesus Die? – Examining the Medical Evidence

- **Station V: Simon of Cyrene Helps Jesus Carry the Cross**
  Location: At the junction of the Via Dolorosa and a side street, where Simon of Cyrene is compelled to carry the cross.

- **Station VI: Veronica Wipes the Face of Jesus**
  This station is where Saint Veronica, moved by the sight of Jesus' suffering, wipes his face with her veil, and his image is miraculously imprinted on it.

- **Station VII: Jesus Falls the Second Time**
  This station marks the second fall of Jesus, which is said to have occurred at the gate leading into the city.

- **Station VIII: Jesus Comforts the Women of Jerusalem**
  This station commemorates Jesus' encounter with the mourning women of Jerusalem, where he tells them to weep not for him but for themselves and their children.

- **Station IX: Jesus Falls the Third Time**
  Near the entrance to the Church of the Holy Sepulchre, this station marks Jesus' third and final fall.

- **Station X: Jesus is Stripped of His Garments**

- **Station XI: Jesus is Nailed to the Cross**

- **Station XII: Jesus Dies on the Cross**

- **Station XIII: Jesus is Taken Down from the Cross**

- **Station XIV: Jesus is Laid in the Tomb**
  These last five stations are all located inside the Church of the Holy Sepulchre, which is believed to encompass both the site of Jesus' crucifixion at Golgotha and his tomb.

Of these traditional stations, three specifically involve Jesus falling:
- Third Station: Jesus falls for the first time.
- Seventh Station: Jesus falls for the second time.
- Ninth Station: Jesus falls for the third time.

The canonical Gospels do not describe these falls, but the Christian church developed a tradition to meditate on the suffering of Christ on his way to crucifixion, which includes these falls.

5. **Noon to 3 PM**:
   - **Crucifixion**: Jesus is crucified alongside two criminals. He speaks several phrases from the cross, including entrusting his mother to the disciple John and asking God why he has been forsaken.
   - **Death**: Jesus dies after several hours on the cross. Notable events at the moment of his death include an earthquake and the tearing of the temple curtain.

Jesus' Exertion on Via Dolorosa

By considering and reevaluating several factors that would have contributed to his physical and emotional strain, we can understand the exertion Jesus experienced on the Via Dolorosa, the path he is believed to have walked to his crucifixion.

Physical Condition Prior to the Walk: Before the journey along the Via Dolorosa, Jesus had already endured a night of trials, abandonment by his disciples, denial by Peter, and the mental anguish of knowing what was to come. Following his condemnation, Jesus suffers physical abuse at the hands of those present. The Gospels describe that Jesus is mocked, struck, and spat upon by some of the temple guards and others present. This abuse likely included slapping or hitting with fists and possibly the use of objects to strike him. Such treatment would have been both physically painful and deeply humiliating.

Additionally, people subjected him to severe physical abuse, which included scourging, a brutal form of Roman punishment that involved using a whip with multiple thongs, often tipped with metal or bone. This would have caused significant blood loss, pain, and weakness.

According to the canonical Gospels of the New Testament, there is no specific mention of the number of lashes Jesus received during his scourging. The Gospels of Matthew, Mark, Luke, and John describe Jesus being scourged or flogged by the Roman soldiers before his crucifixion, but they do not provide a specific number of lashes.

In Jewish law, they could administer a maximum of 40 lashes, but they would deduct one (39 lashes) to prevent accidental violation of the law. Deuteronomy 25:3 describes this practice. However, the Romans, who had the authority to carry out Jesus' scourging, did not have a fixed limit on the number of lashes in such punishments, unlike Jewish law.

There have been various speculations and traditions regarding the number of lashes in Christian tradition and later Christian writings, but these are not based on biblical texts. Christian teachings sometimes mention the number "39 lashes," but this is more likely an inference from Jewish law rather than a historical record of what happened under Roman jurisdiction.

It's important to understand that the Gospels' focus is more on the fact of Jesus' suffering and its theological significance rather than on the specific details of the physical acts. As such, the exact number of lashes Jesus received remains unknown and is not a focal point of the Gospel narratives.

Weight of the Crossbeam: They forced Jesus to carry the patibulum, the horizontal beam of the cross, to the site of the crucifixion. They forced Jesus to carry the patibulum, the horizontal beam of the cross, to the site of the crucifixion.[6] Carrying such a weight, especially after severe physical beating, would have been an immense physical challenge. The act of Jesus carrying the crossbeam (patibulum) to the site of his crucifixion is a significant detail in the narrative of his Passion. Understanding the context and implications of this task further illuminates the extent of the physical challenge it presented.

Before carrying the crossbeam, Jesus had endured severe physical abuse, including the Roman scourging. This brutal whipping would have caused significant blood loss, deep tissue damage, and extreme pain, leaving him in a state of physical shock and weakness. Any additional physical exertion would have been extremely taxing due to the trauma from the scourging alone. While carrying the patibulum, Jesus had to traverse through the streets of Jerusalem to reach his destination. The path may have been uneven, with potential crowds of people. Carrying the heavy beam, especially in his weakened state, would have been a grueling physical ordeal.

The condemned person typically carried the patibulum across their shoulders, bearing the weight on their back and neck. For someone who had just been scourged, the wood rubbing against the open wounds on their back and neck would have caused excruciating pain. In addition to the physical pain, the act of carrying the crossbeam was also a form of humiliation, a public display intended to degrade the condemned individual before execution. The previous night's events – Jesus' arrest, trial, and abandonment by his disciples – compounded his humiliation.

Length and Terrain of the Via Dolorosa: The traditional route is about 600 meters (0.37 miles) long, winding through the streets of Jerusalem. Stone pavements, steps, and slopes, which would have been challenging to navigate while carrying a heavy crossbeam characterize the path, particularly for someone in a weakened state.[7] Understanding its length, terrain, and the condition of the journey provides insight into the physical challenges faced during this pivotal event.

While not an exceptionally long distance in general terms, it would have been an arduous journey under the circumstances of Jesus' physical condition and the weight of the crossbeam. The route winds through the narrow, bustling streets of the Old City of Jerusalem. Their narrowness and winding nature characterizes these streets, typical of ancient city design, making navigation challenging, especially with a cumbersome object like a crossbeam.

Stone slabs are a common feature in ancient cities for paving the path.

---

[6] Hengel, Martin. *Crucifixion in the Ancient World and the Folly of the Message of the Cross*. Fortress Press, 1977.
[7] Murphy-O'Connor, Jerome. *The Holy Land: An Oxford Archaeological Guide from Earliest Times to 1700*. 5th ed., Oxford University Press, 2008.

These stones, worn smooth by centuries of use, could be uneven and slippery, mainly if worn by weather or foot traffic. For someone carrying a heavy burden and in a state of physical exhaustion, maintaining balance and footing on such terrain would add to the difficulty of the journey. The Via Dolorosa includes steps and gentle slopes, given Jerusalem's hilly topography. These elevation changes require additional effort to traverse, especially under the weight of a heavy load. The effort needed to ascend steps or slopes would be significantly taxing for an individual already weakened by physical trauma.

During the time of Jesus, residents, pilgrims, and Roman soldiers would likely have crowded the streets, adding to the complexity of navigating the route. The presence of a large crowd would make the journey physically more challenging and psychologically daunting due to the public nature of the humiliation. The weather and environmental conditions at the time also played a role. Jerusalem in spring can be warm, and physical exertion, combined with potential heat and direct sunlight, would contribute to physical strain and dehydration.

In addition to the physical challenges, the journey bore significant emotional and psychological burden. Onlookers would have witnessed the Via Dolorosa, a physical path intertwined with suffering and humiliation, where some may have jeered and mocked. The journey came to a close at Golgotha (Calvary), the site of the crucifixion. Reaching this destination marked the end of the arduous walk but the beginning of the final stage of Jesus' Passion.

Multiple Falls: According to tradition, Jesus fell three times under the weight of the cross. These falls, depicted in the Stations of the Cross, indicate the extreme physical exhaustion and weakness he was experiencing. These falls do not find explicit mention in the canonical Gospels but have gained significance in the Via Dolorosa narrative within Christian devotional practices.

It is believed that the First Fall (Third Station of the Cross) happens relatively early in the journey. Bearing the weight of the heavy patibulum (crossbeam) and weakened from the severe physical trauma of scourging, Jesus collapses under the cross. This first fall underlines the extreme physical toll that the prior events of his Passion had taken on him. As Jesus continues his journey, he becomes even more weakened and tired, and it is traditionally believed that he has the second fall. This fall might symbolize the ongoing struggle and worsening physical condition as he nears the crucifixion site. It also reflects the relentless nature of the suffering he was enduring.

Many people portray the third and final fall as being close to the destination, Golgotha. At this point, Jesus' physical strength is likely at its lowest, and the weight of the cross, combined with the exhaustion and pain from his injuries, brings him down once more. This fall demonstrates how the physical ordeal reaches its climax, underscoring the severity of his weakened state. The physical act of falling with the weight of the cross would have caused additional pain and injury, exacerbating Jesus' already severe wounds. Emotionally, the experience of falling repeatedly in front of a crowd amidst humiliation and suffering adds another layer to the psychological trauma of the event.

According to the Synoptic Gospels (Matthew, Mark, and Luke), the Roman soldiers compelled Simon of Cyrene to carry the cross for Jesus after one of these falls. This intervention underscores the extreme physical state Jesus was in, unable to take the cross any further. The falls also humanize Jesus, showing his physical limits in the face of overwhelming suffering. Yet, despite these falls, his perseverance in continuing towards Golgotha is seen as a testament to his determination to fulfill his redemptive mission.

Lack of Rest and Sustenance: There is no indication that Jesus was allowed to rest or receive nourishment during this time, which would have further compounded his fatigue and weakness. The lack of rest and sustenance for Jesus during the events leading up to and including his crucifixion is a significant factor that would have significantly exacerbated his physical and mental state.

The events of Jesus' Passion unfolded over an extended period, beginning with the Last Supper on the evening before, followed by the night of prayer and arrest in the Garden of Gethsemane, the series of trials, and finally, the crucifixion. This timeline suggests a prolonged period without any significant rest or sleep. The mental and emotional strain during this period, including the anguish in Gethsemane and the stress of the trials, would have been mentally draining. Stress and anxiety can significantly impact physical stamina and the body's energy reserves.

The physical trauma inflicted, notably the scourging and beatings, would have resulted in substantial blood loss and injury, leading to further depletion of energy. The body's response to injury includes increased metabolic demand, which, without food and water, would lead to rapid exhaustion. There is no record of Jesus eating or drinking from the time of the Last Supper until his crucifixion. This lack of nutritional intake would have meant no replenishment of energy reserves depleted by physical exertion, stress, and trauma.

The lack of fluid intake, blood loss, sweating, and possible exposure to the elements would have led to dehydration. Dehydration alone can cause significant fatigue, weakness, dizziness, and cognitive impairment. Given his weakened state, the effort to carry the crossbeam (patibulum) to Golgotha would have been an enormous physical strain. The lack of rest and sustenance would have made this task even more challenging.

The cumulative effect of sleep deprivation, physical trauma, lack of food and water, and emotional stress would have led to a state of severe physical weakness and exhaustion, making Jesus increasingly vulnerable to the effects of crucifixion. Jesus' strength and commitment to fulfilling his mission are often highlighted in Christian teachings, despite the lack of rest and sustenance that would have made him increasingly vulnerable to the effects of crucifixion during his trials and journey to Golgotha.

Climactic Moment of Suffering: The journey along the Via Dolorosa was the climax of a series of events that involved both physical torture and profound emotional and spiritual anguish. It was not just the physical journey but also the culmination of the path Jesus had been walking throughout his ministry, leading to his sacrificial death. This journey reveals multiple physical, emotional, and spiritual dimensions that intertwine to form the pivotal moment of Jesus' suffering.

By the time Jesus embarked on the Via Dolorosa, he had already endured severe physical torture, including the scourging at the hands of the Roman soldiers, which left him weak and in pain – carrying the heavy crossbeam added to this physical burden. The journey itself, through the narrow and uneven streets of Jerusalem, compounded with the weight of the cross and his weakened state, was a brutal physical ordeal. Emotionally, the Via Dolorosa was a path of profound anguish. Jesus faced not only the physical pain but also the emotional trauma of betrayal, abandonment by his disciples, and the mockery and jeers from the crowds. This public humiliation, coupled with the physical pain, added layers to his suffering.

Theologically, this journey fulfilled the path Jesus had walked throughout his ministry. It was the literal and figurative path to completing his mission – the sacrificial death for the redemption of humanity. Each step on the Via Dolorosa symbolized a move closer to this fulfillment. Via Dolorosa is the pivotal moment in Jesus' life, encompassing the full extent of his physical, emotional, and spiritual suffering. It is the culmination of his earthly ministry and the beginning of a new chapter in the Christian understanding of salvation and redemption.

Chapter 6

Medical Evidence for Trauma

Exploring the wounds of Jesus during the crucifixion from a medical viewpoint requires a careful examination of historical, religious, and medical sources. Although the Christian Gospels are the primary source of information about Jesus' crucifixion, they are not medical records. However, one can analyze the descriptions they offer using modern medical knowledge. It's essential to keep in mind that any description of the wounds involves some degree of speculation since there is no physical evidence or medical records from the time. Nevertheless, we will analyze the Gospel accounts from a medical perspective to determine possible traumatic symptoms that could have led to Jesus' death.

1. Scourging

The scourging of Jesus, as the Gospels describe, plays a significant role in the crucifixion narrative and helps us understand the physical condition Jesus would have been in at the time of his crucifixion. The instrument used for scourging, known as a flagrum, was a Roman whip designed specifically for severe punishment. It typically consisted of several thongs or strands, often made of leather. At the ends of these strands, there were usually metal balls, sharp bone pieces, or hooks. The flagrum was designed in such a way that it caused both striking and tearing of the skin, resulting in deep cuts and bruising.

Tying the person being scourged to a post or a pillar was a common practice, exposing their back, buttocks, and legs. Following that, the Roman soldiers (lictors) would administer the whipping, which was not regulated in terms of the number of strikes. The severity in Jesus' case was determined by the decision of Pontius Pilate, the presiding official. The brutal consequences of such a scourging were evident.

Deep Tissue Damage: The metal or bone at the ends of the flagrum's

Chapter 6 – Medical Evidence for Trauma

thongs would cause deep lacerations, tearing into the skin and underlying muscle tissue. Open wounds significantly increase the risk of infection, especially in unsanitary conditions.[1] While infection would not have been the immediate cause of death in the short timeframe between scourging and crucifixion, it would have contributed to overall weakening.

It's important to keep in mind that sepsis, an infection can cause a potentially life-threatening condition, especially in open wooded areas where infections can spread rapidly. The development of sepsis depends on multiple factors, including the type and source of infection, the individual's overall health, immune system status, and any underlying medical conditions. Although Jesus had no underlying medical condition, the body's response to infection can vary based on several conditions. It's crucial to understand the severity of sepsis and how it can progress quickly.[2]

An infection, which can be bacterial, viral, fungal, or parasitic causes sepsis. If someone has open wounds, it can easily lead to a skin infection and develop into sepsis. The early symptoms of sepsis can appear within a few hours to a few days after the initial infection, such as fever, increased heart rate, increased breathing rate, confusion, or disorientation. Organ dysfunction characterizes sepsis, including decreased urine output, sudden changes in mental status, decreased platelet count, difficulty breathing, abnormal heart pumping function, or abdominal pain. One of the concerning things about sepsis is that it can cause rapid deterioration. Even if someone seems mildly ill at first, they can become critically ill within a matter of hours.

Blood Loss and Shock: The flogging, involving a multi-stranded whip often embedded with metal or bone, would have caused significant cuts and wounds. Such an intense form of physical punishment could result in considerable blood loss, thereby weakening Jesus even before the crucifixion. This substantial blood loss might lead to either anemia, a reduction in red blood cells, or hypovolemia, a decrease in blood plasma volume.

Anemia is a condition characterized by a deficiency in the number or quality of red blood cells or the amount of hemoglobin they contain. This deficiency impairs the oxygen-carrying capacity of the blood. Anemia does not have a strict definition based solely on the volume of blood lost. Instead, it is defined by the impact of that loss (or other factors) on the red blood cells and hemoglobin levels. When blood loss leads to anemia, it's typically because of a decrease in the number of red blood cells, which can happen because of rapid blood loss, such as from a traumatic injury, that can lead to

---

[1] Williams, M. (2021). *Wound infections: an overview.* British Journal of Community Nursing, 26(Sup6), S22-S25. https://doi.org/10.12968/bjcn.2021.26.Sup6.S22

[2] Guarino, M., Perna, B., Cesaro, A. E., Maritati, M., Spampinato, M. D., Contini, C., & De Giorgio, R. (2023). 2023 *Update on Sepsis and Septic Shock in Adult Patients: Management in the Emergency Department.* Journal of Clinical Medicine, 12(9), 3188. https://doi.org/10.3390/jcm12093188

a sudden drop in red blood cell count, resulting in acute anemia.[3] The amount of blood loss that can cause anemia varies widely among individuals and depends on factors like initial health, age, and other medical conditions.

In the case of Jesus, the exact amount of blood loss is unknown. However, experts understand that a reduction in red blood cells would lead to decreased oxygenation, affecting the skin tissue, vital organs, and the brain. The physical strain would become increasingly apparent because of the traumatic beatings, significant blood loss, and extensive deep tissue damage.

Hypovolemia is another condition that develops because of the loss of blood, where the heart is unable to pump sufficient blood to the body, leading to organ failure. Hypovolemia occurs when the heart is unable to pump sufficient blood to the body due to the loss of blood, and the required amount of blood loss can vary depending on factors such as the individual's size, overall health, and the rate of blood loss. Generally, this involves the loss of up to 15% of blood volume, in a typical adult, this would be up to about 750 mL This level of blood loss rarely causes significant changes in heart rate, blood pressure, or mental status in the initial stage.

Some of the common side effects and symptoms of hypovolemia include:

- Dizziness or Lightheadedness: Due to decreased blood flow to the brain.
- Rapid Heart Rate (Tachycardia): The heart compensates for the reduced volume by beating faster.
- Low Blood Pressure (Hypotension): Reduced blood volume leads to a drop in blood pressure.
- Weakness or Fatigue: Caused by inadequate oxygenation and nutrient delivery to the body's tissues.
- Thirst: The body's response to fluid loss.
- Cold, Clammy Skin: because of the constriction of blood vessels and reduced blood flow to the skin.
- Rapid Breathing (Tachypnea): The body attempts to increase oxygen delivery.
- Confusion or Altered Mental State: Resulting from decreased oxygen and nutrient delivery to the brain.
- Pale Skin: because of reduced blood flow and oxygenation.

In severe cases, hypovolemia can lead to shock, a life-threatening condition where the organs do not receive enough blood to function properly. This can cause a range of serious complications, including organ failure.[4]

---

[3] Braunstein, E. M. (2022, September). *Etiology of Anemia*. Johns Hopkins University School of Medicine. Retrieved July 2022, from https://www.merckmanuals.com.

[4] Melendez Rivera JG, Anjum F. *Hypovolemia*. [Updated 2023 Apr 27]. In: StatPearls [Internet]. Treasure Island (FL): StatPearls Publishing; 2024 Jan-. Available from: https://www.ncbi.nlm.nih.gov/books/NBK565845/

Chapter 6 – Medical Evidence for Trauma

Physical Weakness and Exhaustion: The physical trauma resulting from the wounds, coupled with the blood loss, would have led to extreme weakness and fatigue. This condition would significantly heighten the difficulty and pain of carrying the cross to the crucifixion site, as traditionally depicted in the Via Dolorosa. The combined effects of injury and blood loss would heighten the difficulty and pain of carrying the cross to the crucifixion site, making this arduous journey even more grueling.

The severe beating that Jesus endured caused intense pain through several mechanisms. Firstly, the physical impact damages the skin and underlying tissues, leading to inflammation and swelling. This triggers pain receptors in the skin and deeper tissues. When potentially damaging stimuli are present, nociceptors in the skin send signals to the spinal cord and brain. When the skin is injured or exposed to extreme conditions, these nociceptors become activated. They translate the physical stimulus into an electrical signal. Nerve fibers then transmit this signal along to the spinal cord. From there, it travels to the brain, particularly the thalamus and cerebral cortex, where it is perceived as pain. This process not only alerts the body to the presence of harm but also initiates a range of protective responses.[5]

When a person experiences repeated trauma, it can lead to muscle damage and bruising, which can cause further inflammation and pain. The process of inflammation that occurs due to skin injuries involves multiple steps that result in an increase in pain. When the skin is damaged, the damaged cells release certain chemicals such as histamine and bradykinin. These chemicals cause the blood vessels to widen, leading to increased blood flow to the affected area. This, in turn, causes the area to become warm and red. Additionally, the blood vessels become more permeable, allowing the entry of immune cells, oxygen, and nutrients into the tissue. Chemical signals then guided white blood cells to the site of the injury to fight against pathogens.

Another compound produced at the site of damaged tissue and involved in the inflammatory response is Prostaglandins. Prostaglandins are compounds in the body that are crucial in enhancing pain sensation. Tissue damage or infection sites produce prostaglandins, which are compounds that are involved in inflammatory responses. An enzyme called cyclooxygenase (COX) facilitates the production of prostaglandins when cells are injured. These substances contribute to inflammation by causing blood vessels to dilate and increasing the permeability of blood vessel walls. Prostaglandins also sensitize nerve endings, amplifying the pain signals sent to the brain.[6] The combination of tissue damage, inflammation, and the body's pain response results in significant and persistent pain.

---

[5] Dubin AE, Patapoutian A. *Nociceptors: the sensors of the pain pathway*. J Clin Invest. 2010 Nov;120(11):3760-72. doi: 10.1172/JCI42843. Epub 2010 Nov 1. PMID: 21041958; PMCID: PMC2964977.

[6] Landén NX, Li D, Ståhle M. *Transition from inflammation to proliferation: a critical step during wound healing*. Cell Mol Life Sci. 2016 Oct;73(20):3861-85. doi: 10.1007/s00018-016-2268-0. Epub 2016 May 14. PMID: 27180275; PMCID: PMC5021733.

Psychological Effects: Apart from the physical trauma, the process of scourging was also a form of psychological torment. This form of punishment was not only about inflicting physical pain but also about exerting a profound psychological impact on both the victim and onlookers. Victims often spent hours or even days in fear, anticipating the excruciating pain they were about to endure. This anticipation could heighten anxiety and cause immense mental anguish. Performing scourging in public often added a layer of humiliation to the physical pain. This public display served to degrade the victim in front of a community, stripping away their dignity and reducing them to a state of helplessness and vulnerability.

Victims of scourging had no control over the situation, and this helplessness led to feelings of hopelessness and despair. Inability to escape or mitigate the pain only intensified the sense of powerlessness. Amplifying the perception of pain, scourging could have a psychological impact. Prior to the punishment, feelings of fear and anxiety, coupled with shame during the process, could heighten the perception of physical pain.

Individuals who have experienced psychological trauma often report higher pain sensitivity. Trauma can alter the body's pain response system, making a person more susceptible to feeling pain more intensely or more frequently. This phenomenon is sometimes called 'central sensitization,' where the nervous system goes into a state of high reactivity, amplifying pain signals.

Trauma can lead to changes in the brain, particularly in areas that process emotion and pain, such as the amygdala, prefrontal cortex, and the hippocampus. These changes can disrupt how the brain interprets and responds to pain signals, sometimes making the individual more sensitive to physical discomfort or pain. Psychological trauma can lead to psychosomatic pain, where emotional and psychological factors cause physical pain symptoms with no discernible physical cause. This kind of pain is real and can be as debilitating as pain from a physical injury. Trauma can also influence the body's stress response system, particularly the hypothalamic-pituitary-adrenal (HPA) axis, which regulates cortisol production. Chronic stress or trauma can lead to dysregulation of this system, affecting how the body responds to pain. Elevated or depleted cortisol levels can impact pain perception and inflammation.[7]

In summary, the sequence of events leading to the physical and psychological trauma of Jesus began not with the Roman scourging but earlier in the presence of Jewish leaders. As described in Matthew 26:67, temple guards and others subjected Jesus to mockery, physical assault, and spitting, marking the onset of his torturous ordeal. This initial mistreatment in front of Jewish authorities laid the foundation for the subsequent, more severe trauma inflicted by the Romans.

The scourging at the hands of the Romans worsened Jesus' deteriorating

---

[7] Timmers I, Quaedflieg CWEM, Hsu C, Heathcote LC, Rovnaghi CR, Simons LE. *The interaction between stress and chronic pain through the lens of threat learning*. Neurosci Biobehav Rev. 2019 Dec;107:641-655. doi: 10.1016/j.neubiorev.2019.10.007. Epub 2019 Oct 14. PMID: 31622630; PMCID: PMC6914269.

physical and mental state to a significant extent. His immune system was compromised by the deep wounds inflicted by the act of scourging. The release of fluids and antibodies to the damaged tissues initiated a physiological distress response as the body responded to these injuries.

The loss of blood due to scourging had a profound impact on bodily functions. They have reduced blood volume, impaired circulation, and oxygen delivery to vital organs. This lack of sufficient oxygenation, especially to the brain, affected its normal functioning. Jesus' heart rate would have increased to compensate for the reduced oxygen levels, but the impaired lungs struggled with oxygen exchange, leading to systemic exhaustion.

The activation of pain receptors sends continuous electrical signals through the spinal cord to the brain, signaling distress. In response to the intense emotions and physical pain, the brain heightened the pain sensitivity, amplifying the perception and intensity of suffering. At this stage, Jesus' body was highly vulnerable. The continuous loss of blood led to hypovolemia, a state of diminished blood volume, pushing his body toward the brink of sepsis. The open wounds, susceptible to bacterial infection, were further compromised by the unsanitary conditions, undermining the natural defense mechanisms usually provided by intact skin.

By early morning, the physical condition of Jesus had reached a critical point. His ordinarily healthy body was being pushed to its absolute limits, with each minute worsening his dire state.

2. Crown of Thorns

The Crown of Thorns, mentioned in the Gospels, holds great significance in the narrative of Jesus' crucifixion, symbolizing the mockery of his claim to kingship. The crown was probably created using the branches of a thorny plant that grows naturally in the Jerusalem vicinity. The region's flora includes several species with long, sharp thorns, such as the Paliurus spina-christi, Ziziphus spina-christi (also known as Christ's thorn jujube), or acanthus bushes. Several species with long, sharp thorns, such as the Paliurus spina-christi, Ziziphus spina-christi (also known as Christ's thorn jujube), or acanthus bushes, are included in the region's flora.[8] These thorns, often reaching lengths of over an inch, are sturdy and sharp enough to pierce skin. They would have formed a rough circle by weaving or twisting the crown together. The construction method of the crown allowed the thorns to stick out in both directions, ensuring that they would penetrate the scalp when the crown was placed on the head. The crown itself would have been made hastily, reflecting the mockingly cruel intentions of its creators.

Both in its **physical and symbolic implications**, the placement of the crown of thorns on Jesus' head, as described in the Gospels, was a significant act. According to the Gospels, the Roman soldiers placed the crown of thorns on Jesus' head and firmly pressed it down to ensure it stayed in place. This action was a part of the physical torment and a symbolic

---

[8] Dafni, A., Levy, S. & Lev, E. *The ethnobotany of Christ's Thorn Jujube (Ziziphus spina-christi) in Israel.* J Ethnobiology Ethnomedicine 1, 8 (2005). https://doi.org/10.1186/1746-4269-1-8

mockery against His claim of kingship.

The human scalp is among the body's most vascular areas, meaning it has a dense network of blood vessels. This vascularity is necessary to support hair follicles and to regulate temperature. Still, it also means that any injury to the scalp, even a minor one, can result in significant bleeding. While the amount of blood loss from this alone wouldn't typically be life-threatening, it would have been painful and contributed to further weakening Jesus, especially considering the blood loss from the prior scourging.

When the thorns of the crown pierced Jesus' scalp, they would have caused immediate pain due to the abundance of nerve endings in the head. The puncturing of the skin and underlying tissues by the sharp thorns would have resulted in bleeding. Given the number of thorns and the force with which the crown was likely pressed onto his head, the bleeding could have been substantial.

In the context of the time, with a limited understanding of hygiene and infection control, the risk of wound infection would have been high. The open wounds caused by the thorns could quickly have become infected, especially in the unsanitary conditions of a crucifixion site. These infections are typically not specific to the type of thorn but result from bacteria entering the wound. The most common type of infection from a thorn injury is bacterial. Bacteria present on the skin's surface, the thorn itself, or the environment can enter the body through the wound.

Common skin bacteria like Staphylococcus aureus or Streptococcus pyogenes can cause these infections. Puncture wounds, including thorn pricks, can significantly cause tetanus, another severe bacterial infection, especially if the thorns are contaminated with soil or manure. The bacteria produce a toxin that affects the nervous system, leading to muscle stiffness and spasms. Although less common, puncture wounds from thorns can also become infected with fungi present in the environment. This can lead to localized infections or, more rarely, systemic infections.

Beyond the immediate physical suffering it inflicted, this act carried deep **psychological and symbolic significance**, intertwining themes of mockery, humiliation, and a perverted form of coronation that would have profound implications for Jesus' mental state. To understand the psychological impact of the crown of thorns, one must consider the broader spectrum of Jesus' sufferings. Being mocked and humiliated by his captors in the presence of onlookers added a layer of emotional and psychological torment to the physical agony he was enduring. The crown of thorns was not merely a tool of physical torture but a deliberate attempt to belittle and degrade his identity and mission. This act of mockery, aimed at undermining his claims of kingship and divinity, would have been a profound psychological burden, exacerbating the physical pain with the weight of public shame and scorn.

Psychologically, the act of being crowned with thorns and mocked in a twisted coronation ceremony would have been a form of symbolic violence that intensified the emotional and mental anguish. The mocking by Roman soldiers and onlookers, combined with the physical torture, created an environment of total degradation, designed to break down Jesus' psychological resilience and human dignity.

The crown of thorns is a powerful symbol that encapsulates the depth of Jesus' suffering and the profound theological meanings of the crucifixion. It serves as a stark reminder of the physical and psychological torment endured by Jesus but also as a profound symbol of his sacrifice, the nature of his kingship, and the transformative power of suffering and humility. Through this mockery, one could see the true nature of divine kingship, contrasting sharply with worldly notions of power and authority. It invites reflection on the nature of suffering, redemption, and the essence of authentic leadership.

The cumulative effect of the injuries Jesus sustained during his Passion is crucial to understanding the depth and severity of his suffering. The crown of thorns, while not life-threatening on its own, represented a significant addition to a series of brutal physical abuses. These abuses, both in isolation and cumulatively, had profound implications for Jesus' physical state and contributed to the overall trauma he endured. Let's explore these injuries in detail and their combined impact.

The Roman authorities subjected Jesus to scourging, a typical pre-crucifixion punishment, before the crowning with thorns. This process involved being whipped with a flagrum, a whip with multiple leather thongs, often embedded with metal balls or bone pieces. The scourging was intended to weaken the condemned physically to the brink of death without actually killing them, maximizing their suffering on the cross. This brutal beating would have caused deep tissue damage and significant blood loss and left Jesus in a severely weakened state.

The crown of thorns, pressed into Jesus' scalp following the scourging, added to the physical trauma. The scalp is one of the most blood-rich areas of the body, so even superficial puncture wounds could lead to substantial bleeding. This injury, while not fatal, contributed to the overall blood loss and physical degradation, compounding the effects of the scourging.

The cumulative effect of scourging the crown of thorns and carrying the cross contributed to a progressively worsening state of physical trauma. The combined impact of blood loss, shock, dehydration, and physical exhaustion would have significantly accelerated his physical decline. Moreover, the psychological effect of anticipating each subsequent abuse, coupled with the physical pain, would have had a profound impact on Jesus' mental state. The knowledge of impending death, experienced alongside intense physical suffering and public humiliation, constituted a form of psychological torment that compounded the physical agony.

3. Crucifixion

Crucifixion, as employed by the Romans, stood out as a fierce and agonizing form of capital punishment, meticulously designed not merely to bring about death but to do so with maximum pain and humiliation. In the gruesome practice of crucifixion, the executioners meticulously fastened the condemned individual to a large wooden cross, ensuring that the process was as organized as it was merciless. Contrary to the common depictions in art and literature, the nails were not driven through the palms. This misunderstanding overlooks the practical aspect of crucifixion; the human palm is incapable of bearing the weight of the body for an extended period. Instead, they strategically placed the nails through the wrists, allowing the bones to distribute the weight of the body more effectively and preventing the hands from tearing away. Then, they immobilized the victim further by driving a single nail through both feet, affixing them to the vertical stake of the cross. This posture forced the body into an unnatural and strained position, exacerbating the agony of the crucified.

The Romans, understanding the power of spectacle, carried out crucifixions in highly public spaces. The Romans selected these locations deliberately for their visibility, ensuring that the act served as a punishment and a stark warning to others. The crucifixion sites were typically located near busy thoroughfares, outside city walls, or on elevated ground. This strategic placement maximized the public exposure of the condemned, subjecting them to the dual torments of physical pain and profound humiliation. The visibility of the crucified served a dual purpose: it magnified the victim's suffering while also functioning as a powerful deterrent against rebellion or crime, reinforcing Roman authority and the consequences of challenging it.

The methodology and location of crucifixion reveal the calculated cruelty behind this form of capital punishment. It was designed to degrade and dehumanize the victim, inflicting maximum pain and suffering while also serving a sociopolitical purpose. By conducting crucifixions in public spaces, the Romans exploited human suffering as a tool of control, embedding fear in the collective consciousness of the populace. The spectacle of crucifixion, with its deliberate brutality and public humiliation, was a clear message from the authorities, underscoring the cost of defiance and the absolute nature of Roman power.

The method and setting of crucifixion were integral to its function as a form of punishment that went beyond the mere execution of a sentence. They meticulously designed the method and setting of crucifixion to inflict a prolonged and agonizing death, using it as a potent instrument of psychological warfare and societal control. This barbaric yet calculated practice underscored the Romans' mastery of psychological manipulation through the spectacle of suffering, a grim testament to the lengths to which human societies have gone to maintain order and authority.

Physical Effects: Breathing Difficulty

Following a prolonged and arduous ordeal, the physical condition of Jesus deteriorates significantly due to a sequence of violent actions. These include being subjected to physical assault, scourging, and the placement of the crown of thorns, culminating in his crucifixion.[9] The extreme strain

---

[9] Bauman, R. A. (1995). *Crime and Punishment in Ancient Rome*. Routledge.

imposed on his body leads to significant breathing difficulties.

A critical aspect of the agony caused by crucifixion is the physical effects, specifically the breathing difficulties endured by the victim. Due to the victim's suspended position, their respiratory process was severely impaired, causing breathing to become agonizing and ultimately resulting in their death. To understand this further, a detailed medical analysis of the respiratory distress encountered during crucifixion is essential.

When a person undergoes crucifixion, the weight of their body pulling down on the arms and shoulders intensively strains the muscles and ligaments of the chest. This position forces the rib cage into permanent inhalation, making it extremely difficult for the diaphragm to contract and expand generally during breathing. Essentially, the victim is stuck in an inhale position, unable to exhale effectively without physically lifting their body to reduce the tension on the chest muscles and allow the lungs to expel air.[10]

To achieve the necessary lift for exhalation, the crucified individual must push up from the feet and pull up using the arms – actions that cause severe pain due to the nails driven through the wrists and feet. This movement places excruciating stress on the wounds and further traumatizes already damaged tissues and nerves. Additionally, the effort required to lift the body repeatedly to breathe exacerbates muscular fatigue, leading to the rapid onset of exhaustion.

The cycle of attempting to breathe under these conditions leads to a cascade of physiological consequences:

- **Hyperventilation and Respiratory Alkalosis**: The effort to breathe more deeply and the pain associated with each breath can lead to hyperventilation, which may cause respiratory alkalosis. In this condition, the blood becomes too alkaline due to excessive loss of carbon dioxide. This can lead to a series of complications, including muscle twitching and, in severe cases, seizures.

- **Hypoxia and Hypercapnia**: As muscular fatigue sets in and the victim can no longer lift themselves effectively to breathe, oxygen levels in the blood decrease (hypoxia), and carbon dioxide levels increase (hypercapnia). This imbalance further deteriorates the victim's condition, leading to confusion, faintness, and, potentially, loss of consciousness.

- **Acidosis**: Over time, the inability to adequately expel carbon dioxide leads to an acidic environment in the body (acidosis), which worsens the victim's distress and contributes to the failure of various body systems.[11]

---

[10] Zugibe, F. T. (2005). *The Crucifixion of Jesus: A Forensic Inquiry*. M. Evans and Company, Inc.

[11] Hall, J. E., & Guyton, A. C. (2020). *Guyton and Hall Textbook of Medical Physiology* (14th ed.). Elsevier.

The cumulative effect of these physiological stresses – alongside the trauma from other injuries sustained during the crucifixion process – leads to a critical weakening of the body's systems.[12] The heart and lungs are particularly affected, struggling to maintain essential life-sustaining functions amid the deteriorating conditions. Ultimately, the combination of respiratory failure, shock, and cardiac arrest contributes to the victim's death.

The physical effects of crucifixion, especially the breathing difficulties, are a testament to the cruelty of this method of execution. The medical details of respiratory distress during crucifixion highlight not just the physical agony but also the complex interplay of physiological factors that lead to the victim's demise. This analysis underscores the extreme suffering endured and provides a deeper understanding of the ordeal's brutal nature.

Physical Effects: Blood Loss and Shock

The physical effects of crucifixion, particularly the significant blood loss incurred from the nailing of the victim to the cross, are central to understanding the ordeal's physiological impact. When nails pierce the wrists and feet – areas rich in blood vessels and nerves – this invasive trauma initiates considerable hemorrhaging. The resultant blood loss can precipitate a critical medical condition known as hypovolemic shock. This condition, exacerbated by the crucifixion's inherent brutality, unfolds in several stages, each with distinct medical implications.

When nails are driven through the wrists and feet, they likely traverse or come dangerously close to major blood vessels, causing acute bleeding. The wrists, notably, contain the radial and ulnar arteries, and piercing in or near these vessels would lead to substantial blood loss. When injured, the feet's plantar arch, which is similarly vascularized, is susceptible to significant bleeding.[13] The continuous pressure and movement of the body against these nails would increase the wounds' severity and prevent the natural clotting process, exacerbating the blood loss.

Hypovolemic shock occurs when the body loses more than 20% of its blood or fluid supply, making it incapable of ensuring adequate blood flow to the organs. This condition unfolds in several phases:

1. **Compensatory Stage**: Initially, the body responds to decreased blood volume by constricting blood vessels and increasing heart rate, attempting to maintain blood pressure and flow to vital organs. The victim might exhibit symptoms such as anxiety, restlessness, and a rapid but weak pulse.

---

[12] Edwards, W. D., Gabel, W. J., & Hosmer, F. E. (1986). On the Physical Death of Jesus Christ. *Journal of the American Medical Association*, 255(11), 1455-1463.
[13] Kellaway, J. (2003). *The History of Torture and Execution: From Early Civilization through Medieval Times to the Present*. Lyons Press.

Chapter 6 – Medical Evidence for Trauma

2. **Progressive Stage**: As blood loss continues, compensatory mechanisms become overwhelmed. Blood pressure drops sharply, reducing perfusion of organs and leading to altered mental status, pallor, and calm, clammy skin. The reduced oxygen delivery to tissues initiates anaerobic metabolism, causing lactic acid buildup and metabolic acidosis.

3. **Irreversible Stage**: Prolonged inadequate blood flow leads to severe organ damage, particularly to the kidneys, brain, and liver. At this point, even if blood volume is restored, the damage to organ systems may be too extensive for recovery, leading to multi-organ failure and death.[14]

In a modern medical setting, the management of hypovolemic shock involves rapid intravenous administration of fluids and blood products to restore circulating volume, along with measures to control bleeding and support organ function. However, such interventions were not available during crucifixion, leaving the victim to suffer the entire progression of shock with fatal consequences.

The significant blood loss caused by the nailing in crucifixion, combined with the physical and psychological stress of the ordeal, would almost certainly lead to hypovolemic shock. This condition, marked by a critical drop in blood pressure and inadequate blood flow to organs, contributes significantly to the crucifixion's lethality. The interplay between physical trauma, blood loss, and the body's failing compensatory mechanisms highlights the profound suffering and physiological decline experienced during crucifixion.

Physical Effects: Dehydration and Exposure

The physical effects of crucifixion extend beyond the immediate injuries inflicted by nailing and include severe dehydration and exposure to environmental conditions. Victims left hanging on the cross for extended periods – sometimes spanning several days – would inevitably endure the compounded physiological stresses of dehydration, exposure to the elements, and starvation due to the inability to consume food or water. The medical implications of these conditions are profound and contribute significantly to the overall suffering and eventual death of the victim.

Dehydration in crucifixion victims arises from several factors:

- **Lack of Fluid Intake**: The inability to consume water while suspended on the cross directly leads to dehydration.
- **Increased Respiratory Water Loss**: The effortful breathing associated with the crucifixion posture, combined with pain-induced hyperventilation, accelerates water vapor loss through respiration.

---
[14] Ibid 11.

- **Excessive Sweating**: Stress, pain, and heat from direct sunlight increase perspiration, further depleting the body's water reserves.

The physiological consequences of dehydration are multifaceted, with deadly results:
- **Electrolyte Imbalance**: Dehydration disrupts the balance of electrolytes, such as sodium and potassium, which are crucial for nerve and muscle function, potentially leading to muscle cramps, weakness, and arrhythmias.
- **Decreased Blood Volume**: Reduced fluid volume decreases blood pressure and impairs cardiovascular function, exacerbating the effects of blood loss and shock.
- **Renal Failure**: Prolonged dehydration can lead to acute kidney injury as the organs struggle to filter blood and maintain fluid balance.

Exposure to environmental conditions further aggravates the victim's plight:

- **Hyperthermia or Hypothermia**: Depending on the weather, victims could suffer from heat stroke due to direct sunlight and high temperatures or hypothermia from cold winds and nighttime temperatures.
- **Sunburn and Skin Damage**: Prolonged exposure to sunlight can cause severe sunburn, especially in regions with intense UV radiation, contributing to fluid loss and increasing the risk of infection.[15]

The crucifixion's constraints make it impossible for victims to eat or drink, leading to starvation. Deprived of nutrients, the body begins to break down its fat and muscle stores for energy, weakening the victim further and impairing wound healing. The nutritional deficiencies weaken the immune system, increasing susceptibility to infections that could exacerbate open wounds and unsanitary conditions.

The combination of dehydration, exposure, and starvation places immense physiological stress on the body, compounding the effects of injuries and accelerating the decline in bodily functions. These conditions not only intensify the physical and psychological suffering but also significantly contribute to the mechanisms of death in crucifixion, including multi-organ failure and cardiovascular collapse.

The medical details surrounding the effects of dehydration, exposure, and starvation highlight the extreme brutality of crucifixion as a method of execution designed not merely to kill but to inflict maximal suffering over an extended period.

---

[15] Ibid. 11.

Psychological and Emotional Trauma

Apart from physical pain, Jesus would have also been experiencing intense psychological and emotional distress. The humiliation of being paraded through the streets, the hostility of the crowds, and the anticipation of impending death would have been extremely burdensome. The psychological and emotional trauma experienced by Jesus during his journey along the Via Dolorosa and leading up to his crucifixion is a profound aspect of the Passion narrative. This trauma, coupled with the physical pain, contributed to an overall state of extreme distress.

The act of carrying the cross through the streets of Jerusalem was not only a physical burden but also a form of public humiliation. In Roman times, this procession was designed to shame the condemned and serve as a public spectacle. For Jesus, who had a ministry of teaching and healing, this public degradation would have been deeply humiliating. As Jesus was led through the streets, certain individuals in the crowds jeered at him, as described in the Gospels. Being subjected to ridicule and scorn by those he came to teach and save would have been a significant emotional burden.

During this time, Jesus might have felt a profound sense of abandonment. Most of his disciples had fled after his arrest. This isolation amidst a hostile crowd could have contributed to loneliness and despair. The betrayal by Judas Iscariot, one of his disciples, and the denial by Peter would have compounded the emotional trauma. These acts of betrayal by close companions would have been harrowing on a personal and emotional level.

Jesus was acutely aware of the painful and shameful death that awaited him. This anticipation of suffering and the knowledge of his imminent crucifixion would have caused significant emotional distress. As part of his mission, Jesus carried the burden of knowing he was to suffer for the sake of others. This awareness of carrying the sins of humanity and the weight of fulfilling his prophetic mission would have been an immense psychological burden.

The severe physical pain from the scourging, crown of thorns, and carrying the cross would have exacerbated his emotional suffering. Physical trauma can intensify psychological distress, leading to a compounded state of suffering. The Gospels, particularly the account of Jesus praying in the Garden of Gethsemane, suggest that he also experienced spiritual anguish. This anguish likely continued throughout his journey to Golgotha, reflecting a deep inner struggle and distress.

Despite his suffering, accounts of Jesus' journey to crucifixion include moments where he expresses concern for others, such as speaking to the women of Jerusalem. This display of compassion, even amid his torment, highlights the depth of his emotional and psychological strength. The psychological and emotional trauma experienced by Jesus during the events leading up to and including the crucifixion was multifaceted, involving public humiliation, personal betrayal, isolation, anticipation of pain and death, and the spiritual burden of his mission. This aspect of his suffering is a crucial element of the Passion, reflecting his profound anguish.

Chapter 7

Probability of Jesus' Death

Let's delve into the traumatic events that Jesus endured and consider the medical implications of the torture he experienced. By examining the likelihood of death in such a scenario, we can better understand what may have led to the demise of Jesus. However, before we delve into that, let's take a moment to ponder the role of the Roman executioners. Were these individuals amateurs prone to making mistakes, or were they seasoned professionals who knew their job inside and out? The answer to this question may shed some light on the events that took place on that fateful day.

The Roman executioner played a pivotal role in the process of crucifixion, which involved not only physically crucifying the victim but also ensuring the execution was carried out effectively and verifying the death of the condemned. These executioners, often soldiers or professional torturers, possessed extensive knowledge and experience in the methods of torture and execution, making them adept at managing the cruel intricacies of crucifixion. Their responsibilities were multifaceted and required a level of precision and understanding that extended beyond mere physical execution to include aspects of psychological warfare and public deterrence.[1]

The executioners were responsible for the entire crucifixion process, from the initial flogging to the final determination of death. The executioners would typically subject the victim to scourging before crucifying them, which would weaken them through blood loss and pain, making the subsequent crucifixion more unbearable. Executioners would nail or bind the victim's hands and feet to the cross, choosing locations that would maximize pain while prolonging life, such as the wrists and ankles, rather than the palms and feet directly. The positioning ensured that the victim would suffer for an extended period before death, as they calculated. They would affix the victim and then raise and secure the cross in the ground, leaving the victim suspended in an agonizing position.

---

[1] Cook, J. G. (2014). *Crucifixion in the Mediterranean World*. Mohr Siebeck.

Chapter 7 – Probability of Jesus' Death

Determining when a victim had died was a critical aspect of the executioner's role for several reasons:

1. **Ensuring the Sentence Was Carried Out**: Executioners needed to confirm death to fulfill the legal and social mandate of the crucifixion sentence.
2. **Public Deterrence**: The body was often left on the cross to warn others. Ensuring the victim was dead before removal was crucial to this aspect of crucifixion's function as a deterrent against rebellion or criminal activity.
3. **Release of the Body**: In some cases, families were allowed to take the body for burial, but only after death had been unequivocally determined by the executioners.

Roman executioners, tasked with carrying out capital punishments, developed various methods over time to ensure and confirm the death of those condemned.[2] Their techniques were a combination of visual, physical, and sometimes brutal checks, reflecting both the practical needs of confirming death and the often cruel nature of execution practices in ancient Rome:

1. **Lack of Movement**: The most immediate and obvious sign of death was the absence of any movement. Executioners would observe the condemned for a period to ensure there was no muscle twitching or involuntary responses.

2. **Unresponsiveness to Stimuli**: To further verify death, executioners might apply painful stimuli, such as sharp pricks or burns, to see if the condemned reacted. A lack of response indicated that the person's nervous system was no longer functioning.

3. **Relaxation of the Body**: The natural relaxation of the body's muscles after death, leading to a state known as primary flaccidity, was another indicator. This relaxation includes the jaw and limbs, which would visibly droop or sag without any sign of muscle tension.

4. **Changes in Skin Color**: Pallor mortis, the paleness that follows death as blood circulation ceases, was another visual cue used by Roman executioners. This change could happen relatively quickly, within minutes of death.

5. **Breaking the Legs**: In the case of crucifixion, a method designed to be prolonged and excruciating, executioners could hasten death by breaking the condemned's legs. This act, known as crurifragium,

---

[2] Phang, S. E. (2008). *Roman Military Service: Ideologies of Discipline in the Late Republic and Early Principate.* Cambridge University Press.

prevented the individual from pushing themselves up to breathe, leading to asphyxiation. The inability to use their legs to relieve the pressure on their chest and lungs resulted in a quicker death, often from shock or respiratory failure. The lack of reaction to such a severe injury was a clear indicator of death.

6. **Piercing the Side**: As described in the biblical account of Jesus' crucifixion, an executioner thrust a spear into Jesus' side, and witnesses observed the flow of blood and water. This act was a method to confirm death and had physiological implications. The appearance of both blood and a clear fluid suggested a buildup of fluid in the pericardium (the sac around the heart) or pleura (around the lungs), consistent with death. The separation of blood and water could indicate postmortem changes, where blood cells settle under the force of gravity, separating from the plasma (the clear fluid).

This specific detail from the crucifixion narrative highlights the Romans' understanding of death signs, albeit within a religious context. The flow of blood and water would not occur in a living individual in the manner described, suggesting the body had already undergone certain postmortem physiological changes.

7. **Observation of Rigor Mortis**: Although not explicitly described in ancient texts as a method used by Roman executioners, the onset of rigor mortis, the stiffening of the body's muscles hours after death, would be a clear indicator of death if there were any doubts. This process, however, takes several hours to begin and would not be immediate.

The methods employed by Roman executioners reflect a combination of empirical observation and the practical necessities of their role.[3] Execution in Rome was not only a form of punishment but also served a public and political purpose, reinforcing the state's power and serving as a deterrent against crimes. Confirming death was an essential step to ensure the sentence was thoroughly carried out and to prevent any doubt regarding the state's ability to enforce its laws.

The practices of Roman executioners, while brutal by modern standards, were part of a broader cultural and legal system that valued the spectacle of power and the finality of death as a punishment. The detailed methods of confirming death underscore the Romans' practical approach to capital punishment and their understanding of the body's responses to extreme trauma.

---

[3] Naulty, M. (2012). "*The Physiology of Crucifixion*." Journal of the Royal Society of Medicine, 105(4), 115–116.

Cause of Death

There is considerable speculation and various theories regarding the precise cause of Jesus' death. One indisputable fact is that Jesus' heart ceased functioning, ultimately resulting in his death. Based on descriptions of the traumatic events leading to and including the crucifixion, several medical diagnoses could potentially be listed on a death certificate in modern terms.[4] Here are a few possibilities based on the accounts of Jesus' final hours:

A. Cardiac Rupture

The concept of cardiac rupture in the context of intense physical trauma and emotional stress, as speculated in the historical account of Jesus' crucifixion, delves into complex physiological responses under extreme conditions. To understand this further, we can explore the mechanisms and implications of cardiac rupture, pericardial effusion, and hemopericardium in more medical detail.[5]

Cardiac rupture: is a tear in any heart's structures, including the ventricular walls, septum, or papillary muscles. It is most commonly associated with myocardial infarction (heart attack), where a blocked coronary artery leads to the death of heart muscle tissue (necrosis). The necrotic tissue weakens and cannot withstand the pressure from the blood inside the heart, potentially leading to a tear. While less common, intense physical trauma and possibly extreme emotional stress could theoretically contribute to or cause a rupture in a non-atherosclerotic heart through direct or indirect mechanisms, including severe strain on the heart muscle or traumatic injury to the chest.[6]

Pericardial Effusion: This condition involves the accumulation of fluid in the pericardial cavity, the space between the heart and the pericardium (a double-walled sac containing the heart). While small amounts of fluid in this cavity are normal, inflammation, infection, or injury can cause excessive accumulation. In the context of a cardiac rupture, fluid, including blood, can leak into this space, leading to pericardial effusion.[7]

Hemopericardium: Specifically refers to the presence of blood in the pericardial space. This can occur when a cardiac rupture allows blood from the heart chambers to escape into the pericardial cavity. Hemopericardium is a type of pericardial effusion and can rapidly lead to cardiac tamponade. In this critical condition, the pressure from the accumulated fluid prevents the

---

[4] Zugibe, F. T. (2005). *The Crucifixion of Jesus: A Forensic Inquiry*. M. Evans and Company.
[5] Hall, J. E., & Guyton, A. C. (2020). *Guyton and Hall Textbook of Medical Physiology* (14th ed.). Elsevier.
[6] Kumar, V., Abbas, A. K., & Aster, J. C. (2020). *Robbins and Cotran Pathologic Basis of Disease*. Elsevier Health Sciences.
[7] Spodick, D. H. (2003). *Acute cardiac tamponade*. New England Journal of Medicine, 349(7), 684-690.

heart from expanding and filling with blood, severely limiting its pumping ability.[8]

As described, the presence of blood and water (serous fluid) from Jesus' side could theoretically suggest a post-mortem phenomenon where fluid separation occurs within the blood following death, and bodily fluids accumulate due to gravity and lack of circulation. In life, a cardiac rupture involving the heart's right side could allow blood to enter the pericardial space, mixing with the existing pericardial fluid, leading to a mixture of blood and a clear serous component that might resemble the "blood and water" description upon release.

However, it's important to note that in a living person, significant hemopericardium leading to cardiac tamponade would rapidly result in circulatory collapse and death, as the heart could not pump effectively against the pressure of the accumulating fluid. This aligns with the interpretation that the observation of "blood and water" could indicate a fatal event had occurred, consistent with death by cardiac rupture, among other possibilities.

In modern medical practice, diagnosing and confirming cardiac rupture requires imaging techniques such as echocardiography, which can visualize the heart's structures and the pericardial space. Treatment for acute cardiac rupture is emergent surgical repair, while management of pericardial effusion may involve pericardiocentesis (draining the fluid) to relieve pressure on the heart.

The speculation around the cause of death as cardiac rupture, leading to hemopericardium and pericardial effusion, in historical accounts like that of Jesus' crucifixion highlights the intersection of medical science with historical and scriptural narratives. It provides a framework for understanding the physiological processes that can lead to death under extreme duress and trauma.

B. Hypovolemic Shock

The cause of death attributed to hypovolemic shock, in the context of Jesus' crucifixion, is medically plausible given the sequence of events as described historically and scripturally. Hypovolemic shock occurs when there is a significant reduction in blood volume, leading to decreased blood pressure and insufficient blood flow to the organs, ultimately resulting in organ failure and death if not promptly and adequately addressed.[9] The crucifixion and the events leading up to it involved multiple factors that would contribute to such a state.

---

[8] Ibid 7.
[9] Moore, E. E., Feliciano, D. V., & Mattox, K. L. (Editors). (2017). *Trauma*. McGraw-Hill Education.

Mechanisms Leading to Hypovolemic Shock

1. **Severe Scourging**: The process of scourging was a ruthless punishment that involved whipping the back, buttocks, and legs with a flagrum, a whip with several thongs tipped with metal or bone. This would have caused deep lacerations, significant tissue damage, and considerable blood loss, contributing to hypovolemia.
2. **Crown of Thorns and Nailing**: The placement of a crown of thorns on Jesus' head and the nailing of his hands and feet to the cross would have caused further bleeding, both contributing to the overall fluid loss.
3. **Prolonged Crucifixion**: The act of crucifixion itself, a form of execution designed to be protracted and excruciating, would lead to continuous, slow blood loss from the wounds. Additionally, the physical strain and stress would exacerbate the body's metabolic demands, leading to accelerated dehydration and worsening hypovolemic shock.

Pathophysiological impact consists of reduced cardiac output because of significant blood loss, which leads to decreased venous return to the heart and reduced cardiac output. The heart struggles to pump blood effectively, compromising blood flow to vital organs. Initially, the body responds through compensatory mechanisms such as tachycardia (increased heart rate), peripheral vasoconstriction (narrowing the blood vessels to shunt blood to vital organs), and increased breathing rate to improve oxygen delivery. However, these measures are only temporarily effective and can eventually fail as the shock progresses. Reduced blood flow to tissues results in anaerobic metabolism, accumulating lactic acid and metabolic acidosis, further destabilizing the patient's condition. Prolonged hypoperfusion (reduced blood flow) to organs leads to cellular damage and organ failure. Critical organs like the brain, heart, kidneys, and liver are particularly vulnerable.[10]

Given the extent of Jesus' injuries and the physical demands of crucifixion, hypovolemic shock is a medically plausible primary cause of death. The combination of severe blood and fluid loss, stress-induced cardiomyopathy (heart muscle dysfunction due to intense physical and emotional stress), and eventual cardiovascular collapse would lead to a fatal outcome. Cardiac arrest, in this context, would be the final event following a cascade of physiological failures initiated by hypovolemic shock.

In modern medical practice, treatment for hypovolemic shock involves:
- Aggressive fluid resuscitation with crystalloids or blood products.
- Control of the source of bleeding.
- Supportive care to stabilize hemodynamics and organ function.

However, in the historical context of crucifixion, such interventions would

---

[10] Cannon, J. W. (2018). *Pathophysiology of hypovolemic shock*. Critical Care Clinics, 34(1), 43-61.

not have been available, rendering any form of recovery impossible.

This analysis, while speculative in bridging historical and scriptural narratives with modern medical understanding, provides a coherent physiological explanation for the cause of death following the traumatic events described in the crucifixion of Jesus.

C. Asphyxiation

Asphyxiation stands as the most widely acknowledged cause of death in the crucifixion process, a result of the meticulously cruel method designed to prolong suffering and lead to a gradual, agonizing demise. The human body faces a unique set of physiological challenges imposed by the mechanics of crucifixion, with a particular impact on the respiratory system. When the victim is suspended in the position, their ability to breathe normally is severely compromised, which initiates a fatal sequence of events.

A disrupted respiratory function is caused by the crucifixion position, where the body is suspended by the arms in an unnatural posture. Specifically, this posture impedes thoracic expansion and diaphragmatic movement, essential to breathing.

Under normal circumstances, the diaphragm's downward movement and the outward expansion of the rib cage, which creates a negative pressure, drawing air into the lungs facilitates breathing in (inhalation). However, the crucifixion posture forces the rib cage into an expanded state, significantly limiting the ability to inhale effectively. Exhaling (breathing out) in this context becomes an active process requiring the victim to lift their body by pushing up on the feet and pulling on the arms. This movement is necessary to relieve muscle tension and allow the chest to contract to force air out of the lungs.

The cycle of breathing under such conditions is harrowing and physically taxing, leading to several compounding factors that contribute to respiratory failure[11]:

- **Muscle Fatigue**: The effort required to lift the body to breathe places immense strain on the chest, arms, legs, and muscles. Over time, these muscles become exhausted, making it increasingly difficult for the victim to push up to breathe, leading to progressive respiratory failure.
- **Hypoxia and Hypercapnia**: The impaired ability to breathe correctly leads to hypoxia (low oxygen levels in the blood) and hypercapnia (elevated carbon dioxide levels). Hypoxia affects all organs but is particularly damaging to the brain and heart. Hypercapnia, on the other hand, causes respiratory acidosis, further destabilizing the body's acid-base balance and impairing heart function.
- **Acidosis**: The body's struggle to obtain oxygen and expel carbon dioxide leads to metabolic acidosis, where the blood becomes too acidic. This exacerbates the physiological distress experienced by the victim, contributing to the failure of multiple organ systems.

---

[11] Pearn, J. (2014). *Pathophysiology of drowning.* Australian Family Physician, 43(6), 370-373.

Asphyxiation in the context of crucifixion is not merely the result of an inability to breathe; it is a multifaceted process of respiratory failure, exacerbated by physical exhaustion, hypoxia, hypercapnia, and acidosis. These factors, combined with the trauma of crucifixion, including blood loss, dehydration, and shock, create a lethal cocktail that ultimately leads to the victim's death. The medical aspects of asphyxiation during crucifixion underscore the method's brutality, designed not just to kill but to inflict maximum suffering before death.[12]

D. Pulmonary Embolism

Pulmonary embolism (PE) is a potentially life-threatening condition characterized by the sudden blockage of one of the lung's pulmonary arteries, typically caused by blood clots that travel from the legs or other parts of the body (deep vein thrombosis, DVT) to the lungs. In the context of Jesus' crucifixion, several factors related to the event could theoretically increase the risk of developing a PE, leading to a fatal outcome.

Risk factors contributing to PE include[13]:

1. **Prolonged Immobility**: Crucifixion involves being bound or nailed to a cross, which drastically reduces the ability to move. Prolonged immobility is a well-known risk factor for developing DVT, as it leads to blood stasis in the veins, particularly in the lower extremities. Stasis increases the likelihood of clot formation.
2. **Trauma**: The severe physical trauma associated with scourging, nailing, and the stress of crucifixion itself can trigger the body's coagulation cascade, leading to an increased risk of clot formation. Trauma can also cause direct injury to the veins, further predisposing them to clot development.
3. **Dehydration**: The events leading up to and including the crucifixion would likely cause significant dehydration due to both fluid loss from wounds and lack of water intake over an extended period. Dehydration thickens the blood, increasing the risk of clot formation.

When a blood clot lodges in one of the pulmonary arteries, it blocks blood flow to a portion of the lung.[14] This blockage can lead to:

- **Hypoxia**: Reduced blood oxygenation due to impaired gas exchange in the affected lung area.
- **Increased Pulmonary Artery Pressure**: The heart has to work harder to pump blood through the remaining unobstructed lung vessels, which can lead to right ventricular strain and potentially right heart failure.

---

[12] Ibid 4.
[13] Tapson, V.F. (2008). *Acute Pulmonary Embolism*. New England Journal of Medicine, 358(10), 1037-1052.
[14] Kahn, S.R., Lim, W., Dunn, A.S., et al. (2012). *Prevention of VTE in Nonsurgical Patients: Antithrombotic Therapy and Prevention of Thrombosis*, 9th ed: American College of Chest Physicians Evidence-Based Clinical Practice Guidelines. Chest, 141(2_suppl), e195S-e226S.

- **Infarction**: Blockage of blood flow can cause death of lung tissue (pulmonary infarction), leading to further respiratory compromise.

Symptoms of PE can vary widely but often include sudden shortness of breath, chest pain (which may worsen with deep breathing), rapid heart rate, and sometimes coughing up blood. Diagnosis in a modern clinical setting involves imaging studies such as a CT pulmonary angiography (CTPA) to visualize the clot or D-dimer blood tests to detect fragments of clot dissolution. However, these diagnostic tools would not have been available in ancient times.

In the case of Jesus' crucifixion, a pulmonary embolism could theoretically arise as a complication of the crucifixion process, given the risk factors present. The sudden onset of respiratory distress, rapid heart rate, and potential hypoxia leading to collapse and death could be consistent with a large or saddle PE, which is a large clot blocking the main artery of the lung or branches right and left, causing immediate and severe hemodynamic instability.

However, it's important to note that while PE could be a contributing factor, the multiple traumas and physiological stresses associated with crucifixion make it likely that we cannot definitively pinpoint a single cause of death. Pulmonary embolism, if it occurred, would be one of several critical factors leading to the fatal outcome, alongside hypovolemic shock, cardiac rupture, or other complications related to the severe physical trauma and execution method.

E. Acute Respiratory Distress Syndrome (ARDS)

Acute Respiratory Distress Syndrome (ARDS) is a severe lung condition characterized by the rapid onset of lung inflammation. Fluid causes it accumulating in the tiny, elastic air sacs (alveoli) in the lungs, which prevents the lungs from filling with air and, consequently, decreases oxygen levels in the bloodstream.[15] ARDS can result from various direct and indirect causes of lung injury, including severe trauma and shock, both of which could be associated with the events leading to Jesus' death.

The leading causes of ARDS in the context of crucifixion are[16]:

1. **Severe Trauma**: The physical trauma from scourging, nailing, and the crucifixion itself can initiate systemic inflammatory responses, which could contribute to lung injury and the development of ARDS.
2. **Infection**: Open wounds and the potential for infection could lead to sepsis, another common cause of ARDS. Sepsis induces a systemic inflammatory response that can severely damage lung tissue.

---

[15] Matthay, M.A., Zemans, R.L., Zimmerman, G.A., et al. (2019). *Acute Respiratory Distress Syndrome.* Nature Reviews Disease Primers, 5(1), 18.
[16] Ware, L.B., & Matthay, M.A. (2000). *The Acute Respiratory Distress Syndrome.* New England Journal of Medicine, 342(18), 1334-1349.

3. **Shock**: The profound shock resulting from extensive blood loss and emotional distress could compromise blood flow to the lungs and other organs, further exacerbating lung injury.

Symptoms of ARDS develop quickly, often within hours or days following the injury or the onset of the inciting illness, and include:
- Severe shortness of breath.
- Labored and extremely rapid breathing.
- Hypoxemia (low blood oxygen levels), despite receiving oxygen.
- Confusion and extreme fatigue due to decreased oxygen delivery to the brain.

Diagnosis of ARDS in a modern clinical setting is based on clinical findings and imaging tests, such as chest X-rays or CT scans, which show widespread infiltrates across both lungs, indicative of edema, without evidence of heart failure (which can cause similar symptoms).

Given the severe physical trauma and physiological stress associated with crucifixion, ARDS represents a plausible contributing factor to the cause of death. The intense inflammatory response triggered by the trauma and subsequent shock could lead to fluid accumulation in the alveoli, severely impairing gas exchange. This would result in critical hypoxemia and respiratory failure, conditions that, without modern medical interventions like mechanical ventilation and supportive care, would likely be fatal.

While it's challenging to attribute Jesus' death to a single medical condition due to the complex interplay of severe trauma, shock, and potential complications such as infection, ARDS could undoubtedly have been a significant contributing factor, leading to respiratory failure and, ultimately, death.

F. Cardiac Arrest Due to Physical and Emotional Stress

The concept of cardiac arrest because of a combination of severe physical injury, loss of blood, dehydration, and extreme emotional stress encompasses a complex interplay of physiological mechanisms that can lead to the heart's inability to pump blood effectively. In the context of Jesus' crucifixion, this explanation incorporates multiple contributing factors that could culminate in cardiac arrest.

Cardiac arrest occurs when the heart suddenly stops beating due to a malfunction in its electrical system, leading to a cessation of blood flow to the brain, lungs, and other organs. It's a critical emergency that, if not treated immediately, results in death. Contributing factors in the context of crucifixion includes[17]:

1. **Severe Physical Injury**: The crucifixion and events leading up to it, including scourging, would cause significant physical trauma. Such

---

[17] Myerburg, R.J., & Junttila, M.J. (2012). *Sudden Cardiac Death Caused by Coronary Heart Disease*. Circulation, 125(8), 1043-1052.

trauma triggers a systemic inflammatory response, releasing various mediators that can have direct toxic effects on the heart muscle, potentially leading to myocardial depression and dysfunction.

2. **Loss of Blood**: Significant injury loss reduces the blood volume available to circulate, leading to decreased oxygen delivery to tissues, including the heart muscle. The heart must work harder to circulate the reduced volume of blood, which can stress an already weakened or damaged heart.

3. **Dehydration**: Resulting from blood loss and potentially limited fluid intake before and during the crucifixion, dehydration would further concentrate the blood, increasing its viscosity and making it harder for the heart to pump. Dehydration also leads to electrolyte imbalances, disrupting the heart's electrical activity.

4. **Extreme Emotional Stress**: Psychological stress can have profound effects on the heart. Acute emotional stress can lead to a surge in adrenaline and other stress hormones, increasing heart rate and blood pressure and demanding more oxygen from the heart muscle. In extreme cases, this can lead to stress-induced cardiomyopathy, also known as "broken heart syndrome," where part of the heart temporarily enlarges and doesn't pump well.

The combination of these factors can lead to several potential mechanisms for cardiac arrest, including[18]:

- **Electrolyte Imbalances**: Crucial for heart muscle contractions and electrical conduction, imbalances can lead to arrhythmias.
- **Hypoxia**: Reduced oxygen delivery to the heart muscle impairs its function and can lead to fatal arrhythmias.
- **Acidosis**: Resulting from severe blood loss and tissue hypoxia, acidosis can impair myocardial function and disrupt the heart's electrical stability.
- **Cardiomyopathy**: Severe stress and physical trauma can directly damage the heart muscle, reducing its ability to pump effectively.

In the absence of modern medical intervention, the combination of these factors would likely result in a cascade leading to cardiac arrest. The severe physical stress of crucifixion, compounded by blood loss, dehydration, and emotional stress, could overwhelm the heart's ability to function correctly. Arrhythmias induced by electrolyte imbalances, hypoxia, acidosis, or acute heart failure from myocardial damage could all lead to the sudden cessation of effective heart pumping (cardiac arrest).

---

[18] Wittstein, I.S., Thiemann, D.R., Lima, J.A., et al. (2005). *Neurohumoral Features of Myocardial Stunning Due to Sudden Emotional Stress*. New England Journal of Medicine, 352(6), 539-548.

Understanding cardiac arrest in this context provides insight into how a culmination of extreme conditions and physiological stressors can lead to the heart's failure to maintain its critical role in circulation, ultimately resulting in death. While speculative in bridging historical events with modern medical understanding, this explanation aligns with known medical principles regarding the body's response to extreme stress, injury, and physiological compromise.

While modern medical diagnostics can offer these potential explanations, the historical and scriptural accounts provide limited details. Thus, these theories are speculative and aim to understand the physiological processes that could have led to Jesus' death, considering the traumatic events described.

CHAPTER 8

CONFIRMATION OF JESUS' DEATH

In the previous chapter, we delved into the crucial role of the Roman executioner in confirming the death of anyone under their responsibility. According to the text, which describes the crucifixion and the treatment of Jesus' body, there is medical evidence indicating that Jesus was pronounced dead when he was removed from the cross. By analyzing this evidence, we can gain a deeper understanding of the significance of this event.

The Spear Wound

Based on historical accounts of the crucifixion, the wound inflicted by a spear to the chest could provide a plausible explanation for the observed outflow of "blood and water." The cardiovascular and thoracic anatomy of the chest area and the body's physiological responses to trauma can help us understand this phenomenon. The chest cavity contains vital structures, including the heart, which is enclosed in the pericardial sac, and the lungs, which are surrounded by the pleural cavities.[1] If a spear wound were to penetrate this area, it could potentially damage major blood vessels, the heart, and also rupture the pleural or pericardial spaces.

In the unfortunate event that a spear pierces through the human form to strike the heart, the aftermath unfolds in a sequence of devastating events, painting a grim tableau of survival hanging by a thread. Such an assault on the heart is not merely a breach of flesh but an invasion into the sanctum that orchestrates the rhythm of life itself. The spear, upon penetrating the heart or its significant vessels – the aorta, the grand conduit carrying oxygen-rich blood from the heart to nourish the body, or the vena cava, the large vein tasked with returning oxygen-poor blood to the heart for rejuvenation – unleashes a catastrophic cascade of bleeding. This bleeding is not a silent

---

[1] Moore, K.L., Dalley, A.F., & Agur, A.M.R. (2018). *Clinically Oriented Anatomy*. Wolters Kluwer.

Chapter 8 – Confirmation of Jesus' Death

affair; driven by the heart's relentless pumping, blood is forced out through the wound with every beat, a grim rhythm marking the loss of life force with each pulsation.

Yet, the tragedy does not end with the outward flow of blood. The pericardium, a double-layered envelope, encases the heart and serves as a physical barrier and a limit to the heart's motion, ensuring it remains positioned within the chest. When the spear's intrusion extends to this critical membrane, it compromises the pericardium's integrity, leading to a condition known as hemopericardium. This term describes the accumulation of blood within the pericardial cavity, a scenario as dire as the external bleeding, if not more so. With blood collecting in this confined space, pressure mounts against the heart, constraining its ability to expand fully with each beat. This restricted movement hampers the heart's capacity to pump effectively, threatening to disrupt the delicate balance of blood circulation that sustains life.

In its relentless duty, the heart becomes entangled in a vicious cycle: it attempts to circulate blood while spilling and containing its lifeblood, ultimately resulting in a dangerous decline in its function. The scenario is a poignant reminder of the heart's vulnerability and the intricate dependencies that sustain human life. In this detailed narrative of destruction, the spear does not merely wound flesh – it unravels the threads of existence, challenging the heart's resilience and the body's capacity to endure.

In the dire circumstances of crucifixion, a harrowing form of punishment, the human body experiences extreme stress, dehydration, and the exacerbation of any pre-existing conditions, such as heart failure. This crucible of suffering can lead to fluid accumulation in spaces not typically found in such volumes, specifically within the pleural cavity that houses the lungs and the pericardial cavity cradling the heart. These accumulations, known as pleural and pericardial effusion, are akin to the body's distress signals, evidence of its struggle under the weight of its torment.[2]

The pleural effusion involves the collection of fluid in the pleural space, the slender slit between the lung's outer surface and the inner chest wall. Under normal circumstances, this fluid serves as a lubricant, allowing the lungs to expand and contract smoothly with each breath. However, trauma, such as crucifixion or inflammation from various causes, disturbs this delicate balance. The effusion can compromise breathing, encasing the lungs in a suffocating embrace, making each breath labor.

Similarly, pericardial effusion, fluid accumulation in the pericardial cavity, signifies distress in the heart's vicinity. This cavity, designed to facilitate the heart's movements, can become a trap when filled with excess fluid, exerting pressure on the heart, hindering its ability to pump blood efficiently, and potentially leading to a dire condition known as cardiac tamponade.

The origins of these effusions, the "water" component of the fluids, are

---

[2] Karmy-Jones, R., Nathens, A., & Jurkovich, G.J. (2004). *Management of traumatic lung injury: A Western Trauma Association Multicenter review*. Journal of Trauma and Acute Care Surgery, 57(6), 1343-1349.

multifaceted, rooted in the body's response to the extreme stress of crucifixion, dehydration, and the aggravation of existing ailments like heart failure. These conditions collectively contribute to the body's precarious state, manifesting as pleural and pericardial effusions. Such accumulations of fluid are more than mere symptoms; they are the silent witnesses to the body's fight for survival against the overwhelming odds imposed by such a brutal method of execution.

In the unfortunate event where a chest trauma, such as a spear wound, pierces through to the heart's vicinity, it sets off a catastrophic chain reaction within the body. The injury can cause bleeding into the pericardial space, the slender, fluid-filled gap between the heart and its protective sac, the pericardium. Under normal circumstances, this space contains just enough fluid to allow the heart to beat freely and smoothly within the confines of the chest cavity. However, the consequences are dire when an influx of blood from the trauma disrupts this delicate balance.

This accumulation of blood in the pericardial space is the harbinger of a condition known as cardiac tamponade. The term itself paints a picture of distress, where the excess fluid exerts pressure on the heart, squeezing or "tamponading" it. As the volume of blood increases in this confined space, it begins to compress the heart, restricting its muscular walls from expanding fully. This compression is not merely a mechanical inconvenience; it is a life-threatening impediment. The heart's chambers, particularly the ventricles, cannot fill to their usual capacity during diastole, the cardiac cycle phase, when the heart relaxes and fills with blood.

The ripple effects of this compromised filling are profound. When less blood enters the heart, it pumps out even less to the body with each contraction. This reduction in cardiac output – the amount of blood the heart pumps per minute – can rapidly decline circulatory function. Blood pressure plummets, organs begin to starve for oxygen, and without prompt and effective intervention, the body can spiral into circulatory collapse.

Cardiac tamponade, therefore, is not just a response to chest trauma but a critical emergency that underscores the delicate balance upon which life hangs. In its intricately connected way, the body signals the urgency of the situation through symptoms like rapid heart rate, low blood pressure, and signs of shock. It's a stark reminder of how a single moment of trauma can unravel the complex and finely tuned system that is the human body, bringing to the forefront the fragility of life in the face of such overwhelming odds.

The integrity of the chest is compromised by a traumatic event, such as the violent penetration of a spear, causing the repercussions to extend far beyond the initial wound. In its destructive path, the spear may breach the pleural cavities, the slender spaces that envelop the lungs in a protective embrace. This invasion can result in significant bleeding into these cavities, heralding the onset of conditions known as hemothorax and hemopneumothorax, each marking a grave turn in the victim's physiological response to the chest trauma.

Hemothorax, the accumulation of blood within the pleural cavity, emerges as blood escapes from the damaged vessels and fills the space between the lung and the chest wall.[3] This accumulation can have a dual detrimental effect: it not only deprives the lung of its necessary room to expand but also exerts pressure on the lung tissue, effectively collapsing the lung to a degree. The act of breathing, so vital and yet so often taken for granted, becomes an uphill task, with each breath bringing scant oxygen into the body, insufficient to meet its demands.

Compounding this peril is the possibility of a hemopneumothorax, characterized by blood and air in the pleural cavity. Introducing air into this space, typically a vacuum that enables the lungs' expansion and contraction, disrupts the negative pressure essential for normal lung function. The combined presence of air and blood within the pleural cavity exacerbates the lung's collapse and further impedes its ability to function. This condition not only complicates the process of oxygen exchange but also places the individual in immediate danger of respiratory failure.

The body's response to these conditions is a testament to the severity of the situation. As the affected lung struggles to fulfill its role in oxygenating the blood, symptoms such as rapid breathing, chest pain, and a noticeable decrease in oxygen saturation levels manifest, signaling the body's distress. The physiological struggle to maintain adequate oxygen levels becomes an uphill battle, with every breath a reminder of the delicate balance disrupted by the trauma.

In the narrative of survival against chest trauma, hemothorax and hemopneumothorax stand as critical chapters, marking the body's resilience and vulnerability in the face of grave injury. These conditions not only underscore the immediate impact of the trauma but also the intricate interplay between structure and function within the human body, where the breach of a physical barrier can lead to profound physiological consequences.

When a severe wound, such as a spear piercing the chest, is inflicted on the human body, the clinical implications become both immediate and dire. The first and most palpable consequence of this traumatic injury is the rapid loss of blood. This hemorrhage is not just a simple leakage; it is a torrential outpouring of life's essence from the circulatory system, where blood volume and pressure are critical for sustaining life. The breach in a major blood vessel, especially as pivotal as the aorta – the main artery responsible for distributing oxygen-rich blood from the heart to the rest of the body – or the heart itself, sets the stage for a catastrophic loss of blood volume.

This swift depletion of blood volume precipitates a state of shock, a condition characterized by the failure of the circulatory system to provide adequate blood flow to the organs and tissues of the body. Shock is not merely a single event but a progressive crisis; as blood pressure plummets, the perfusion of organs is severely compromised, leading to cellular damage and the potential for multiple organ failure. In an attempt to preserve blood flow to vital organs, the body constricts blood vessels in less critical areas, a

---

[3] Kulshrestha, P., Munshi, I., & Wait, R. (2004). *Profile of chest trauma in a level I trauma center*. Journal of Trauma, 57(3), 576-581.

compensatory mechanism that can only sustain life for a brief window of time without medical intervention.

The specter of immediate death looms large in such scenarios, significantly if the wound directly impacts the heart or aorta. Damage to the heart disrupts its ability to pump blood effectively, quickly becoming incompatible with life. Similarly, a wound to the aorta is tantamount to a death sentence without swift and decisive medical intervention, given the aorta's critical role in the circulatory system. In such cases, the margin between life and death is razor-thin, hinging on the location and severity of the injury as well as the immediacy and effectiveness of the emergency response.

In the clinical narrative of a spear wound penetrating the chest, the immediate consequences sketch a grim picture of the fragility of human life when confronted with such violent trauma. This scenario underscores the precarious balance between life and death, where the swift loss of blood can usher in a speedy end, highlighting the critical need for urgent and skilled medical intervention to forestall the direst outcomes.

In the aftermath of a traumatic chest wound inflicted by a spear, assuming the victim miraculously escapes the clutches of immediate death, their battle is far from over. The secondary effects that follow, such as cardiac tamponade, hemothorax, or hemopneumothorax, cast long shadows over their chances of survival. These conditions, each a formidable adversary in its own right, share a common thread in their potential to be fatal without swift and effective medical intervention – an option regrettably not available in times past.

Cardiac tamponade, where blood accumulates in the pericardial space, exerting pressure on the heart, can quickly lead to heart failure. This condition demands urgent drainage of the accumulated blood to relieve pressure on the heart and restore its pumping function. Without such intervention, the heart cannot maintain adequate blood circulation, and the body's tissues and organs begin to suffer from oxygen starvation, a condition leading inexorably toward death.

Similarly, hemothorax, the presence of blood in the pleural cavity, compromises lung function by preventing the lungs from fully expanding. This results in inadequate oxygenation of the blood, a critical issue that, if not addressed through procedures like thoracentesis to remove the blood, can result in respiratory failure and death.

Hemopneumothorax introduces an even more complex challenge, with blood and air filling the pleural space. This condition impedes lung expansion and disrupts the negative pressure vital for lung function. The combined effect of impaired oxygen exchange and reduced lung volume necessitates immediate chest tube insertion to evacuate the air and blood, a procedure crucial for survival but unavailable in historical contexts.

In the era before modern medicine, the grim reality was that such secondary effects of chest trauma often spelled a death sentence. The absence of contemporary medical interventions, such as surgical drainage, thoracentesis, or chest tube insertion, meant that the physiological derangements induced by these conditions went unchecked. The body, despite its remarkable capacity for self-healing, could not overcome the mechanical and physiological barriers imposed by these sequelae of chest trauma. Thus, in the shadow of a spear wound, the specter of death loomed large, not just in the immediate aftermath but in the gradual, inevitable decline brought on by conditions that we now, with the benefit of medical advancements, can often successfully treat.

Conclusion – The phenomenon of "blood and water" flowing from a spear wound to the chest, as described in historical accounts of the crucifixion, finds its explanation within the realms of medical science. This occurrence, far from being a mere detail, offers profound insights into the devastating impact of such an injury. We can understand the "blood and water" as a mixture of hemorrhagic fluids from cardiovascular injuries and the clear serous fluid from pleural or pericardial effusions. This detail not only corroborates the lethal nature of the wound but also highlights the intricate physiological processes at play.[4]

The initial outpouring of blood signifies the immediate and severe compromise of the cardiovascular system, potentially involving significant vessels or the heart itself. This bleeding, if unchecked, rapidly spirals into hemorrhagic shock, a condition marked by the critical reduction in blood volume and the subsequent failure of the circulatory system to sustain life. Concurrently, the presence of "water," or serous fluid, points towards the secondary effects of the trauma, such as the accumulation of fluid in the pleural cavity (hemothorax or hemopneumothorax) or around the heart (cardiac tamponade). These conditions exacerbate the dire situation by compromising respiratory functions or directly impairing cardiac efficiency.

Therefore, one cannot overstate the gravity of such a wound. It presents a multifaceted threat to life, combining the immediate peril of hemorrhagic shock with the looming dangers of cardiac tamponade and respiratory compromise. This lethal trifecta underscores the swift and merciless path to death that such an injury could pave, absent the interventions of modern medical science.

In delving into the anatomy and physiology underlying such traumatic injuries, a coherent medical framework emerges that not only sheds light on the mechanisms of death in crucifixion but also enriches our understanding of historical narratives. The reported observation of "blood and water" flowing from a spear wound is thus not merely an anecdotal detail but a window into the brutal reality of crucifixion, offering a vivid testament to the severity of the act and the acute physiological responses it elicits. This insight bridges the gap between historical accounts and contemporary medical understanding. It provides a holistic view of the physical toll exacted by crucifixion and the indelible mark it leaves on the annals of human history.

---

[4] Zugibe, F.T. (2005). *The Crucifixion of Jesus: A Forensic Inquiry.* M. Evans and Company.

Lack of Movement

The complete halt of both voluntary actions, such as deliberate movement, and involuntary responses, like reflex actions, serves as a primary indicator of death, marking the total cessation of bodily functions. This absolute quiescence stems from the cessation of the nervous system's operations and the complete absence of brain activity, which is responsible for controlling both conscious and subconscious movements.[5] Exploring the medical dimensions of this occurrence offers a deeper understanding of the mechanisms that culminate in the body's ultimate immobility and highlights the significance of these signs in determining death.

The cerebral cortex, a sophisticated brain region tasked with conscious thought, decision-making, and deliberate action choreographs the intricate dance of voluntary movements. With its vast networks of neurons, this cerebral command center orchestrates every deliberate gesture, from the simple act of waving a hand to the complex coordination required for speech. However, with death's onset, this vibrant activity hub falls silent. The cessation of life heralds the total loss of cerebral function, effectively extinguishing the spark that fuels our capacity for voluntary actions.

When the cerebral cortex stops functioning, it loses its control over voluntary movements, making the body unable to initiate intentional actions. This profound stillness is not merely the absence of motion but a stark testament to the cessation of the brain's commanding role. Without the electrical impulses and neural commands that once surged through the cortex, directing every conscious movement, the body enters a state of complete inactivity. This transition into motionlessness marks a definitive end to the body's interaction with the world through voluntary movements, encapsulating the finality of death in the cessation of the cerebral cortex's function.[6]

The phenomenon of voluntary movement, a capability we often take for granted, is underpinned by an intricate symphony of communication within the body. This symphony involves a complex network of signals that flow between the brain, the spinal cord, and the muscles, all orchestrated by the somatic nervous system. This system serves as the conduit for our intentions to be translated into physical action, enabling us to interact with the world around us through movement. The seamless collaboration between our neural infrastructure and our muscular system is at the heart of this process. This partnership allows for the precise execution of movements ranging from the mundane to the marvelously complex.

However, death irreversibly disrupts the delicate balance of this system. As life ceases, so does the flow of neural signals that facilitate voluntary movement. The interruption of communication between the brain, the spinal cord, and the muscles marks a definitive end to the body's ability to move at

---

[5] Purves, D., Augustine, G. J., Fitzpatrick, D., et al. (2020). *Neuroscience*. Sinauer Associates.
[6] Wijdicks, E. F. M. (2002). *The diagnosis of brain death*. New England Journal of Medicine, 344(16), 1215-1221.

will. The interruption of communication between the brain, the spinal cord, and the muscles definitively ends the body's ability to move at will, as it no longer receives electrical impulses to direct muscle contraction or relaxation in response to conscious desires.

This cessation of voluntary movement is a stark reminder of the vitality of the somatic nervous system's role in our daily lives. The silence that follows its shutdown is profound, signaling the end of the body's dynamic interaction with the world. In this state of final immobility, the intricate ballet of neural signals and muscular responses that once allowed for voluntary actions concludes, illustrating the irreversible nature of death and the comprehensive cessation of bodily functions that accompany it.[7]

The autonomic nervous system orchestrates countless involuntary movements that persist without our conscious command, weaving the tapestry of life. This remarkable system is an unseen conductor, ensuring the heart's steady beat, guiding the smooth, wave-like contractions that propel food through the intestines, and overseeing the occasional, involuntary twitch of muscle fibers. It operates in the shadow of our awareness, diligently regulating the essential functions underpinning our existence.

Yet, with the advent of death, this continuous stream of autonomous activity comes to a halt. Ceasing its operations, the autonomic nervous system, once a tireless guardian of the body's internal processes. Once the heart, whose beats were as reliable as the rising sun, falls silent. In the midst of its movement, the intricate dance of peristalsis, which moves sustenance through the digestive tract, ceases. Even the slightest muscle twitches, those spontaneous signs of life, fade away.

This cessation of autonomic functions marks a profound transition. The system that operated beyond the reach of our will, that quietly maintained the rhythms of life, withdraws its influence. With its departure, the body relinquishes the last vestiges of motion, entering a state of eternal stillness. This moment underscores the definitive end of the biological processes that once animated the physical form, sealing the conclusion of life's complex ballet of involuntary movements.

Reflexes, those instinctive, automatic movements that our bodies execute in direct response to certain stimuli, are the most basic form of involuntary action governed by the spinal cord and the brainstem. These actions, such as the swift contraction of the pupil in bright light or the involuntary gag when something touches the back of the throat, are our body's immediate defenses, requiring no conscious thought to activate. They are the body's primal responses, hardwired into our nervous system, ensuring protection and essential physiological functions without cognitive intervention.

However, the journey into death brings about a profound silence in these once vigilant areas of the central nervous system. The absence of reflexive actions – a pupil that no longer contracts in response to light, a throat that

---

[7] Goldstein, D. S. (2011). *Principles of Autonomic Medicine*. National Institutes of Health.

does not gag at the unexpected intrusion – marks a significant milestone in the cessation of life. This lack of reflex signifies that the spinal cord and brainstem, the centers responsible for mediating these automatic responses, have ceased their activity.

The cessation of reflexive actions heralds a profound shift, indicating that the most fundamental layers of the body's nervous system have stilled. It underscores the irreversible departure from life, where the fundamental mechanisms that once offered protection and automatic regulation no longer respond to the external world. This moment captures the completeness of death's embrace, as even the body's most instinctual functions fall quiet, leaving behind the silent testimony of a life that once was.

In the quiet aftermath of death, the human body embarks on a transformative journey through various stages, one of the most recognizable being rigor mortis. This phenomenon, rigor mortis, manifests as a gradual stiffening of the muscles, a stark testament to the profound changes unfolding within the muscle fibers at a chemical level. It marks a departure from the living state, where muscles contract and relax with ease, to a phase of temporary immobility.

The onset of rigor mortis is not immediate but begins several hours post-mortem, peaking in intensity before gradually dissipating over a period that can extend up to 72 hours.[8] The cessation of the body's biochemical machinery, which continuously pumps energy into muscle cells to facilitate contraction in life, is the underlying cause of this stiffening. In death, as these processes grind to a halt, calcium ions accumulate within the muscle fibers, prompting them to contract in a way that no longer reverses without the energy-rich molecule ATP, which is no longer produced.

Contrary to what might be assumed, the rigidity that observers notice during rigor mortis does not indicate lingering life or active muscle movement. Instead, it is a clear indicator of the opposite – a body in transition, moving through the post-mortem stages dictated by the natural laws of biology and chemistry. Rigor mortis stands as a silent marker of time. This phase encapsulates the body's final farewell to the dynamism of life before the eventual relaxation that follows as part of the decomposition process.

As the journey beyond death continues, the body reveals another hallmark of its new state: livor mortis. This phenomenon emerges as a visible sign, where gravity pulls the blood to the body's lower regions, manifesting as a purplish discoloration on the skin. This change is not merely cosmetic but speaks volumes about the cessation of life's vital functions. It signifies that once the relentless engine drives blood circulation, the heart has stilled its beats forever, allowing the blood to pool where gravity dictates.

Livor mortis begins to take shape a few hours after death, becoming most pronounced within 12 to 24 hours before gradually becoming fixed, painting a stark picture of the body's finality. This blood settling, far from a simple change, underscores the profound stillness that has overtaken the body.

---

[8] Madea, B. (2016). *Handbook of Forensic Medicine*. Wiley-Blackwell.

Chapter 8 – Confirmation of Jesus' Death

With the heart's silence, the dynamic flow of blood that once coursed through veins and arteries, delivering oxygen and nutrients to every cell, ceases. The body, no longer governed by the pulsing rhythm of life, submits to the immutable pull of gravity, marking the skin with the telltale signs of livor mortis.

This phase of post-mortem change offers a tangible reminder of the body's transition from a state of vibrant activity to immutable stillness. Livor mortis, with its distinct discoloration, is a testament to the cessation of the heart's function and the end of blood circulation, serving as a visual cue to the irreversible passage from life to death.

Within the confines of a clinical setting, medical professionals approach determining death with a meticulous blend of observation and technology. Medical professionals undergo training to search for unequivocal signs that indicate the cessation of life. These signs include the stillness of the body, characterized by the absence of both voluntary movements, which are consciously controlled, and involuntary movements, which occur beyond our conscious control. However, the process extends beyond mere observation, incorporating the absence of heartbeat, the halt of respiration, and the silence of brain activity as critical indicators of death.

Modern medicine uses sophisticated tools designed to peer into the body's complex systems with precision to navigate this solemn task. Electrocardiograms (ECGs) stand at the forefront of this endeavor, offering a window into the heart's activity – or lack thereof. By monitoring the electrical signals that orchestrate the heartbeat, ECGs provide a definitive account of the heart's status, capturing when its rhythmic beats cease.

Parallel in importance is the electroencephalogram (EEG), a tool that measures the whispers of electrical activity within the brain. In life, the brain is a hub of ceaseless activity, with EEGs tracing the undulating waves of brain signals. In death, these waves fall silent, and once a testament to the brain's vibrant activity, the EEG line flattens, signaling the cessation of cerebral functions.

The determination of death, therefore, is a confluence of clinical acumen and technological insight. It involves a comprehensive assessment that not only observes the physical stillness of the body but also delves into the electrical silence of the heart and brain. Through this rigorous process, we ensure that the declaration of death is based on unequivocal evidence, marking the transition from life to an irreversible stillness with clarity and precision.

Conclusion – The unmistakable sight of a body in utter stillness, devoid of any movement, whether by conscious will or automatic function, is a profound testament to the cessation of life. This profound stillness is not merely a physical state but a marker of the complete halt in the brain and nervous system's activities, signifying that the intricate machinery of the body has come to a standstill. Within the context of the Gospel narratives, which depict Jesus' body as devoid of motion, there is a resonant echo of these medical truths. Through their portrayal of motionlessness, these ancient texts tacitly affirm the reality of death, aligning with the contemporary medical understanding that associates such stillness with the cessation of all bodily functions.

Delving into the physiological underpinnings of this state of stillness sheds light on the mechanisms that signal the end of life, offering a bridge between the observable signs of death and the internal cessation of vital activities. This exploration not only validates the criteria used across ages to determine death but also deepens our understanding of how death is identified and confirmed in historical and modern medical practices. By understanding the cessation of movement as a universal indicator of death, we gain insight into the continuity of medical observation, from the narratives of ancient texts to today's clinical settings, underscoring the timeless nature of this ultimate human transition.

Absence of Breathing

Breathing, an involuntary, essential rhythm of life, ceases at death as a definitive sign that life has ended. Governed by the brainstem, this critical function relies on the harmonious work of respiratory muscles, orchestrating the vital exchange of gases that sustains life: drawing oxygen into the lungs and expelling carbon dioxide.[9] In the grim scenario of crucifixion, a method of execution that often leads to death by asphyxiation, the halt of visible breathing is not merely a cessation of an autonomic process but a stark marker of life's final phase.

The medical intricacies of breathing involve the dynamic interplay between neurological control and muscular action, ensuring that the lungs continuously maintain their capacity to facilitate gas exchange. However, asphyxiation disrupts this delicate balance, gradually diminishing the body's ability to oxygenate blood and remove carbon dioxide, leading to critical levels of oxygen deprivation and carbon dioxide accumulation. This state of hypoxia and hypercapnia, respectively, triggers a cascade of physiological failures, culminating in the cessation of heart function and, ultimately, all bodily processes.

The absence of breathing in the context of crucifixion starkly illustrates the end of the struggle for life, highlighting asphyxiation as the mechanism through which death ensues. This cessation, observable as the body's failure to draw breath, offers a clear, unequivocal indication of death, aligning with medical understanding of the vital role breathing plays in sustaining life. By exploring the mechanisms underpinning breathing, the devastating effects of asphyxiation, and the signs that signify life's end, we gain a deeper appreciation for the complexities of this fundamental process and the profound implications of its loss.

A complex system of neural regulation nestled within the brainstem governs breathing, an essential lifeline. This area, known as the respiratory center, serves as the command hub for the autonomic control of respiration,

---

[9] West, J. B. (2012). *Respiratory Physiology: The Essentials*. Wolters Kluwer Health.

Chapter 8 – Confirmation of Jesus' Death

orchestrating the rhythm of life without conscious effort from us. It meticulously monitors the body's oxygen and carbon dioxide levels, the critical gases in the delicate balance of life, and adjusts the rate and depth of breathing accordingly.[10]

This intricate process ensures that the respiratory center responds with precision when the body requires more oxygen, such as during physical exertion, or needs to expel an excess of carbon dioxide. It signals the respiratory muscles to increase the pace and depth of breaths, facilitating a greater intake of oxygen and a more rapid expulsion of carbon dioxide. Conversely, when the body's demand for oxygen decreases or the levels of carbon dioxide are low, the respiratory center dials back the effort, reducing the rate and depth of breathing.

Through this automatic, behind-the-scenes adjustment, the body's remarkable capacity for self-regulation ensures that the vital exchange of gases in the lungs is maintained in accordance with its fluctuating needs. Through this seamless integration of neural commands and muscular action, breathing mechanisms sustain the continuous flow of life, underpinning our very existence with each breath taken.

Breathing is an intricate process that involves a harmonious interplay of muscles and structures within our body, enabling us to take in oxygen and expel carbon dioxide. This process begins with inhalation, a phase where a remarkable muscular action takes place. At the heart of this action is the diaphragm, a large, dome-shaped muscle at the base of the lungs. When we inhale, the diaphragm contracts and moves downward, expanding the space in the chest cavity. This movement is not solitary; it is accompanied by the intercostal muscles located between the ribs. These muscles engage and expand the rib cage outward, further increasing the volume of the chest cavity.

This increase in chest volume during inhalation creates a negative pressure differential between the inside of the lungs and the outside atmosphere. It's akin to pulling back on a syringe plunger while the nozzle is blocked; the internal vacuum created desperately needs to be filled. In the case of breathing, this vacuum draw air through the nose or mouth, down the trachea, and into the lungs, filling the microscopic alveoli where gas exchange occurs.

Exhalation, however, tells a different story, especially when the body is at rest. This phase is predominantly passive and a natural consequence of relaxing muscles. The diaphragm relinquishes its contracted state, moving upwards into its dome shape, and the intercostal muscles ease, allowing the rib cage to contract. This relaxation reduces the volume of the chest cavity, increasing the pressure inside the lungs compared to the outside air. This pressure gradient pushes the air out of the lungs, expelling the carbon dioxide-rich air back through the trachea and out of the body.

However, this narrative of exhalation can change under certain conditions, such as stress or during vigorous exercise. In these situations,

---

[10] Feldman, J. L., Del Negro, C. A., & Gray, P. A. (2013).

exhalation evolves from a passive to an active process. Additional muscles, particularly those in the abdomen and the inner intercostal muscles, come into play. These muscles forcefully decrease the volume of the chest cavity, pushing air out of the lungs with more power to accommodate the increased demand for oxygen exchange during these intense periods.

This continuous cycle of inhalation and exhalation, governed by the rhythmic muscular action of the diaphragm and intercostal muscles, illustrates the body's remarkable ability to adapt to its needs, ensuring that every cell receives the oxygen it requires sustaining life and expels the carbon dioxide produced as a waste product.

Crucifixion, an ancient method of execution, imposes a cruel burden on the human body, particularly affecting the vital process of breathing. Executioners meticulously design the position in which victims are placed during crucifixion to induce a slow, agonizing death, with asphyxiation being a critical factor in this lethal equation. The mechanics of breathing, so effortlessly carried out under normal circumstances, become a Herculean task for the crucified.[11]

In this torturous position, the body's weight pulls downwards, causing the arms to stretch out and the chest to expand in a fixed, unnatural state. This severely compromises the ability to utilize the diaphragm and intercostal muscles – those critical components in breathing. The diaphragm, unable to contract and descend effectively, and the intercostal muscles, strained beyond their capacity to aid in expanding and contracting the rib cage, leave the victim struggling for every breath.

To draw air into their lungs, victims must make a desperate, painful effort to lift their bodies. This act requires pushing up from the feet, often pierced by a nail, and pulling on the arms, extending the already overstretched muscles and bones. Each breath demands a physical battle against gravity and the body's injuries, a battle that becomes increasingly insurmountable as fatigue and exhaustion set in.

As time passes, the victim's ability to lift themselves to breathe diminishes, causing longer intervals between breaths and a progressive decrease in the amount of air they can draw in. This vicious cycle leads to a critical reduction in oxygen levels in the blood and an accumulation of carbon dioxide, edging the victim closer to the brink of asphyxiation.

This method of execution, therefore, not only inflicts physical pain through the initial act of crucifixion but also enforces a prolonged struggle for breath, culminating in a slow, suffocating death. The tragic irony lies in the body's mechanisms for survival becoming the instrument of its demise, as the natural act of breathing transforms into an exhaustive ordeal, ultimately leading to the victim's asphyxiation.

The harrowing ordeal of crucifixion pushes the human body into a dire state of distress, marked by a gradual yet relentless decline in the essential function of breathing. This decline triggers a cascade of physiological crises, beginning with hypoxia, a condition characterized by dangerously low levels

---

[11] Pearn, J. (2014). *Pathophysiology of drowning.* Australian Family Physician, 43(6), 370-373.

Chapter 8 – Confirmation of Jesus' Death

of oxygen in the blood. The crucified individual, struggling to breathe, is unable to lift their body adequately to draw in air, causing the oxygen supply to their tissues and organs to dwindle and disrupting the delicate balance necessary for life.

The immediate consequence of this oxygen deprivation is hypercapnia, an accumulation of carbon dioxide in the bloodstream. Under normal circumstances, each exhalation expels carbon dioxide from the body. However, the impaired respiratory efforts in crucifixion hinder this process, causing carbon dioxide to build up. This excess carbon dioxide in the blood leads to a drop in pH, a state known as acidosis, where the blood becomes too acidic. Acidosis further impairs cellular functions and can damage vital organs, exacerbating the body's distress.

This toxic combination of hypoxia, hypercapnia, and acidosis initiates a domino effect of bodily failures. The heart, brain, and other organs, starved of oxygen and overwhelmed by acidity, begin to malfunction. This organ failure is not instantaneous but results from the prolonged and compounding effects of the body's inability to maintain its essential functions under the strain of crucifixion.

As the body's systems falter, the victim edges closer to death. The tragic trajectory from the initial difficulty in breathing to the onset of hypoxia, followed by the accumulation of carbon dioxide leading to acidosis, and culminating in the body's shutdown's organs, outlines a brutal process that underscores the cruelty of crucifixion as a method of execution. It is a slow, agonizing descent into darkness, marked by the body's gradual surrender to the inescapable grip of asphyxiation.

In the quiet moments of observation, when the breath draws its final, unseen passage through the body, the most telling sign of life's cessation becomes starkly evident through the stillness of the chest and abdomen. Under the watchful eyes of those seeking to discern the line between life and death, the absence of visible movement in these regions speaks volumes. Typically, the chest rises and falls with each breath, a rhythmic dance of life that sustains us from our first cry to our last sigh.[12] The abdomen, too, plays its part, subtly expanding and contracting in concert with the chest, a visible testament to the body's ceaseless effort to draw in oxygen and expel carbon dioxide.

Yet, in the somber reality where breathing ceases, this movement halts. The chest no longer rises, nor does the abdomen swell with the tide of breath. This stillness, this profound absence of the natural undulation that signifies air exchange, marks a critical threshold. It signals that the vital act of breathing, the essence of life, has stopped. Once a dynamic symphony of involuntary movements meant to sustain life, the body now lies silent, devoid of the breath that animates the soul.

This cessation of movement is not merely the absence of a single breath but a definitive indicator that the pathways of air, once bustling with the flow of life, have stilled. It's a silent testament to the end of the body's struggle for

---

[12] Saukko, P., & Knight, B. (2016). *Knight's Forensic Pathology*. CRC Press.

oxygen, a sign that the heart and lungs have relinquished their tireless rhythm. Observing this stillness, the absence of chest or abdominal movement becomes a poignant marker, a final note in the symphony of life that, once ceased, signifies the solemn moment of departure from the living world.

In the meticulous examination of signs of life, especially in the quiet, anticipatory space between hope and finality, using a stethoscope becomes a bridge to the body's inner workings. This instrument, designed to magnify the whispers of our physiology, becomes pivotal in assessing the presence or absence of breath sounds. Under normal circumstances, breathing generates a symphony of sounds: the gentle rush of air during inhalation, a testament to the lungs filling with life-sustaining oxygen, and the soft whoosh of exhalation, marking the body's release of carbon dioxide.

However, the stethoscope reveals a starkly different narrative in the critical assessment for the cessation of breathing. An ominous silence prevails when placed against the chest or back, where the lungs should reveal the tell-tale signs of their function. This absence of sound, the lack of any auditory evidence of air moving through the trachea and bronchial tubes, confirms a profound and final truth: there is no airflow through the respiratory tract.

This silence detected through the stethoscope is not merely the absence of noise; it is a definitive indicator that the vital exchange of gases, the essence of the respiratory process, has ceased. It signifies that the lungs, those resilient organs tasked with exchanging life-giving oxygen and waste carbon dioxide, no longer carry out their sacred duty. Therefore, the absence of breath sounds signifies a crucial and undeniable marker that the spark of life, carried on the wings of breath, has been extinguished. In the delicate balance between life and death, this silent testimony captured through the stethoscope is a solemn confirmation of life's departure.

In the unfolding narrative of determining the cessation of breathing, the body itself can manifest visible cues, among which cyanosis stands out as a poignant marker. Cyanosis, characterized by a bluish skin discoloration, emerges as a visible testament to a critical shortfall in oxygen within the bloodstream. This condition paints its mark most noticeably around the lips and fingertips, where the thinness of tissue layers starkly reveals the underlying blood's hue. Under the specter of ceased breathing, this bluish tint emerges as a dire signal, illustrating the profound impact of oxygen deprivation on the body's physiology.

Cyanosis develops as oxygen levels plummet, unfolding a visual narrative of distress that leaves behind a tell-tale sign that can be both observed and interpreted. As blood circulates without sufficient oxygen, it adopts a darker coloration, a visual cue that casts a bluish shadow over life's palette when seen through the skin. This discoloration is a critical indicator that the mechanisms of breathing, the processes that oxygenate the blood and sustain life, may have ground to a halt.

However, the tale of cyanosis is nuanced, woven with threads of caution. While this bluish discoloration starkly signals a severe lack of oxygen, it does

Chapter 8 – Confirmation of Jesus' Death

not, in isolation, narrate the end of life's story. Cyanosis can also manifest in the living, a grim herald of severe hypoxia, where the body, still clinging to life, struggles against the suffocating grip of oxygen scarcity. This sign, indicative of the dire state of oxygenation, must be read in conjunction with other signs to discern the boundary between life's struggle and its cessation.

Thus, cyanosis stands as a visible marker on the body's surface, a signal of the silent turmoil within, indicative of the cessation of breathing yet calling for a careful interpretation in the broader context of assessing life's presence or its departure.

In contemporary medicine, medical professionals approach the solemn task of confirming death with a meticulous and multi-faceted strategy. This process extends beyond observing breath cessation, incorporating clinical signs and sophisticated diagnostic tools to ascertain the irreversible halt of life's functions. The journey to this final confirmation weaves through the basic tenets of physical examination and advanced medical technology.

At the forefront of this examination is the assessment of several fundamental indicators of life. Medical professionals seek the pulse, the rhythmic beating of the heart that sends waves of blood through the body's vessels, a primary signal of life's persistence. The absence of this pulse speaks volumes, indicating the stillness of the heart. Furthermore, medical professionals assess the body's responsiveness to external stimuli, and an unresponsive state further suggests the departure of life. Additionally, the medical professionals gauge the eyes through the pupillary light reflex test, shining a light into the eyes to check if the pupils contract, indicating active neural pathways. The absence of this reflex further corroborates the cessation of vital activities.

Venturing into the domain of advanced diagnostics, electrocardiography (ECG) emerges as a critical tool. This technique allows clinicians to trace the heart's electrical activity, searching for the rhythmic patterns that denote a beating heart. An ECG can reveal silence where the electrical symphony of cardiac life should be, offering a clear, graphic representation of the heart's status. Similarly, capnography, which measures the concentration of carbon dioxide in expired air, gives a window into the respiratory system's function. In life, each exhalation carries away carbon dioxide, a byproduct of cellular metabolism. As shown by capnography, the absence of this gaseous exchange underscores the cessation of respiratory activity.

Together, these techniques and observations form a comprehensive approach to confirming death. By employing both the senses and technology, this process respects the complexity of human life and ensures that the declaration of death is made with certainty. This careful, respectful confirmation process underscores the gravity of the moment, marking the final chapter in the story of a life with dignity and precision.

Conclusion – In the context of the crucifixion, the narrative that unfolds around Jesus' final moments is deeply intertwined with the clinical manifestations of death by asphyxiation – a fate all too common in this ancient and brutal method of execution. The physical limitations placed on the victim, notably the severe restriction of respiratory movements, set the stage for a gradual but inevitable progression towards asphyxiation. This physiological ordeal, marked by the inability to draw breath due to the body's strained position effectively, mirrors the grim reality faced by those subjected to crucifixion.

As the accounts detail Jesus ceasing to breathe, this pivotal moment is not just a marker of his suffering but also a critical juncture that aligns with medical interpretations of death's approach. The cessation of breathing, a fundamental life-sustaining process, is a stark indicator that the bodily functions essential for life are failing. While significant, this observation gains further weight when analyzed through the lens of contemporary medical practice, which advocates for a holistic approach to confirming death. This approach entails recognizing the absence of breathing and corroborating this with other signs, such as the lack of a pulse, unresponsiveness to external stimuli, and the absence of neurological reflexes like the pupillary light response.

Moreover, applying advanced diagnostic techniques, such as electrocardiography (ECG), to assess heart activity and capnography to evaluate respiratory function enriches this confirmation process. These methods provide tangible evidence of the cessation of cardiac and respiratory activities, offering a scientific basis to support the clinical determination of death.

Thus, the historical accounts of the crucifixion, particularly the observation of Jesus' final breaths, are not only a narrative of a poignant and transformative moment but also a scenario that resonates with the physiological and clinical understanding of death by asphyxiation. This convergence of historical narrative and medical insight sheds light on the physiological realities underpinning the crucifixion, offering a comprehensive view that bridges the gap between ancient practice and modern medical principles. It underscores the clinical finality of death as both a physiological and a medically verifiable event, grounding the historical accounts of crucifixion in the realities of human biology and medical science.

No Response to Stimuli

Within the realms of medical science and forensic examination, the body's reaction – or lack thereof – to external stimuli stands as a cornerstone in the definitive determination of death. This principle serves as a critical demarcation line, distinguishing between the profound depths of unconsciousness or coma, where life still flickers within, and the absolute stillness of death. The narratives chronicled within the Gospel accounts place a poignant emphasis on Jesus' absence of reflexive movement or reaction, notably when someone prodded him or when a spear was thrust into his side. This stark unresponsiveness highlights unequivocal evidence of his demise.

Delving this aspect from a medical vantage point necessitates a deeper exploration of the body's innate reflexive actions. These reflexes – automatic

Chapter 8 – Confirmation of Jesus' Death

responses to external stimuli – are fundamental neurological and physiological function markers. They are the body's automatic way of responding to the environment, safeguarding itself against harm. These reflexes signal the functioning of the brain and central nervous system, underlying the complex interplay of life processes within.[13]

However, the absence of these reflexive responses opens a window into the cessation of the body's vital functions. In clinical and forensic contexts, this absence is not merely a passive observation but a significant indicator that the intricate systems that govern life have stopped. When stimuli elicit no response, it suggests a breakdown in the communication channels between the brain, the nervous system, and the body at large – a hallmark of death's finality.

In the context of the Gospel narratives, the detailed accounts of Jesus' lack of reaction to being prodded and the spear wound are profound illustrations of this principle. These moments are not just narrative details but critical observations, aligning with medical understandings of death. The absence of any reflexive response to such invasive stimuli underscores the conclusion that vital functions have ceased. This convergence of historical documentation and medical insight provides a compelling framework for understanding death. It emphasizes the significance of the absence of reflexive responses as a clear, medically grounded indicator of the cessation of life, offering a nuanced perspective on the events described in the Gospel accounts through the lens of medical science.

A suite of automatic reactions that operate beneath conscious awareness is governed by the spinal cord and the brain's more primal regions in the human body, which is equipped with a remarkable array of reflexive responses. These reflexes serve as the body's immediate defense mechanisms, reacting to external stimuli quickly and precisely, independent of our conscious decision-making processes. Among these are the instinctive withdrawal from sources of pain, the rapid blink reflex triggered by objects suddenly nearing the eyes, and the involuntary cough or gag reflex that activates when the throat encounters an unexpected stimulus.

Even when the veil of unconsciousness envelops an individual, these reflexive actions remain vigilant guardians of the body's well-being. Their presence is a testament to the activity within the brainstem, the crucial nexus of nerve connections that bridge the brain and spinal cord. This area of the brain, even in states of deep unconsciousness, continues to orchestrate these fundamental responses, offering evidence of its unyielding operation.

These reflexes are not just simple bodily quirks but are vital indicators of the brainstem's health and, by extension, the overall vitality of the central nervous system. The ability to elicit such reflexes in an unconscious individual reveals a still-sparking ember of life within, signaling that the core systems responsible for life's most basic functions remain active. This reflexive repertoire, from flinching away from harm to blinking away threats and rejecting harmful substances through coughing or gagging, underscores

---

[13] Ibid 6.

the body's innate capacity to protect and sustain itself, even in the absence of conscious control.

Within the medical assessment of an individual's fundamental physiological condition, doctors observe reflexive responses alongside a constellation of vital signs. These vital signs, comprising the heartbeat, breathing patterns, blood pressure, and the body's reaction to stimuli, serve as the pillars of health monitoring, painting a comprehensive picture of a person's immediate state of being. Each component offers a glimpse into the human body's intricate workings, revealing its life-sustaining systems' efficiency and resilience.[14]

The heartbeat, with its rhythmic pulsation, is a testament to the enduring efforts of the heart as it diligently pumps blood throughout the body, delivering oxygen and vital nutrients to tissues and organs. Breathing, an equally rhythmic and essential life process, ensures oxygen exchange for carbon dioxide, a critical transaction that sustains cellular function and overall vitality. Blood pressure, the force exerted by circulating blood on the walls of blood vessels, offers insight into the dynamic interplay between the volume of blood the heart pumps and the resistance of the blood vessels, a balance crucial for maintaining life.

When assessing the response to stimuli, medical practitioners delve into neurological function, probing the depths of consciousness and the brainstem's activity. This evaluation seeks to uncover the presence or absence of reflexive responses, actions that, despite their simplicity, speak volumes about the brain's capacity to perceive and react to the environment. The lack of such responses, particularly in contexts where they should be readily elicited, suggests a profound disruption in brainstem activity, a scenario that often accompanies the gravest of outcomes.

The absence of a heartbeat and respiration stands as the most stark and undeniable indicators of death. When extinguished, these vital signs, so fundamental to life, signify the cessation of the body's most basic functions.[15] A further lack of response to stimuli cements this conclusion, providing unequivocal evidence that the neurological pathways, once bustling with activity, have fallen silent. This holistic approach to monitoring vital signs and reflexive responses thus offers a detailed and nuanced understanding of an individual's physiological state, allowing medical professionals to discern the delicate boundary between life and the cessation of all vital activity.

In the realm of medical diagnosis and the solemn task of determining death, the absence of any response to painful stimuli stands as a profound indication of a deep cessation within the body's core functions. This unyielding silence in the face of stimuli that would, under normal circumstances, provoke an immediate and involuntary reaction, points to the brain's relinquishment of its most fundamental roles. Such a lack of response signals that the intricate web of neural activity, which orchestrates everything from thought to the basic instinct to withdraw from harm, has stilled.

---

[14] Swearingen, P. L. (2016). *All-In-One Nursing Care Planning Resource*. Elsevier.
[15] Wijdicks, E. F. M. (2001). Determining brain death in adults. *Neurology*, 56(10), 1005-1015.

Chapter 8 – Confirmation of Jesus' Death

The confirmation of the cessation of life is not made in isolation, but is guided by medical professionals who weave this critical observation into a broader tapestry of clinical signs. Reflex actions hard-wired into the very architecture of our nervous system, stand among the body's most resilient functions. Their persistence, even as consciousness fades, is a testament to their deep-seated role in human physiology. Therefore, when these reflexes fade into stillness, the implication is clear and stark: the brain, the command center for these reflexes, has ceased its activity.

The medical implications of such a finding are profound. In the delicate balance between life and death, the presence or absence of these reflexive responses provides crucial insight into the state of the brain's activity. Their disappearance marks not just a failure of the body's ability to react to immediate threats but signifies a deeper, more irreversible cessation. It's a point of no return, where the absence of these last bastions of bodily autonomy confirms beyond doubt that life's journey has ended.

These understandings in medical practice shape the approach to diagnosing death, ensuring that the declaration is made with certainty and sensitivity. The convergence of these observations – particularly the lack of response to painful stimuli with other indicators of life's cessation – forms the foundation upon which the final confirmation of death rests, a testament to the meticulous and compassionate nature of medical science in facing the ultimate threshold of human existence.[16]

In the clinical environment, where healthcare professionals navigate the boundaries between life and death with precision and care, the confirmation of death holds significant weight when reflexive responses are absent. When observed alongside the unmistakable cessation of heartbeat and the stillness of breath, this silent testimony forms a triad of indicators that unequivocally signals the end of life. These primary observations, deeply rooted in the physiological essence of living, offer a clear, initial confirmation of death, marking a moment of profound significance.

However, the journey to this final determination continues beyond observing these physical signs alone. The field of medicine, with its ever-advancing technologies and methodologies, extends the capacity to confirm death with an even greater depth of certainty. Electroencephalography (EEG), a sophisticated technique that measures the brain's electric activity, emerges as a pivotal tool in this process. By tracing the brain's electrical patterns, EEG provides a window into the ongoing activity – or lack thereof – within the brain's complex networks. The absence of electrical activity on an EEG is a powerful indicator that the brain, the seat of consciousness and control, has ceased its functions.

Further enhancing the diagnostic arsenal are various imaging studies, which allow clinicians to visually assess the brain and other vital organs for signs of life. These technologies, ranging from computed tomography (CT) scans to magnetic resonance imaging (MRI), offer detailed insights into the body's internal state. They can reveal the structural integrity of the brain and

---

[16] Ibid 13.

other organs and any residual activities that might escape more conventional assessments.

These advanced diagnostic techniques bolster the initial clinical observations, providing a comprehensive and nuanced understanding of the body's state. The integration of EEG findings and imaging studies with the absence of reflexive responses, heartbeat, and breathing weaves a conclusive narrative of death. This multifaceted approach reflects the complexity of life itself and the meticulous care with which medicine approaches its cessation, ensuring that the declaration of death is made with the utmost accuracy and respect for the individual.

In the epochs before the advent of modern medical technology, the most fundamental observations guided ancient civilizations to discern the boundary between life and death. Without the sophisticated instruments today's medical practitioners have, these early observers turned their attention to the readily apparent signs of life's presence or its departure. The cessation of movement, the stillness of breath, and an unyielding lack of response to external stimuli were their compass in navigating the solemn territory of death determination.

The Gospel accounts, along with other historical narratives, exemplify this reliance on observable phenomena and provide a poignant illustration of these ancient practices in action. The specific detail that Jesus showed no reaction to the spear thrust – a moment laden with significance – resonates deeply with the methods employed by those in antiquity to confirm death. Such an act, invasive and undeniably painful, would, under normal circumstances, provoke a reflexive response in a living individual. Therefore, the absence of such a response was a powerful testament to the cessation of life, aligning with the time-honored criteria for recognizing death.

This episode chronicled within the Gospels, provides a narrative account of a moment in history and offers insight into the continuity of human experience in grappling with the reality of death. It underscores how, across millennia, humans have sought to understand and mark the transition from life to death, relying on the most elemental signs when more complex measures were beyond reach. This ancient practice, mirrored in the Gospel account of Jesus' final moments, bridges the gap between past and present, highlighting a fundamental aspect of the human condition: the quest to comprehend and acknowledge the cessation of life with the means available from the simplest observations to the most advanced technologies.

In forensic science, investigating the mysteries of death extends beyond the mere fact of its occurrence to probe the specifics of its timing. In scenarios where the arsenal of modern technology is not at hand or when the circumstances of death demand a return to foundational principles, examining the body's interaction with external stimuli becomes a crucial element of inquiry. This investigative approach hinges on assessing whether the body exhibits reflexive or responsive actions when stimulated, a critical clue in the puzzle of death's chronology.

This method, rooted in the basic tenets of human physiology, is further enriched by observing post-mortem changes, offering a more

Chapter 8 – Confirmation of Jesus' Death

comprehensive picture of death's aftermath. Among these, rigor mortis – the stiffening of the muscles that occurs several hours after death – and livor mortis – the pooling of blood in the lowermost parts of the body, resulting in a purplish discoloration of the skin – are pivotal. These phenomena unfold in a relatively predictable sequence following death, providing forensic experts with tangible markers to estimate the time since death occurred.

The interplay between the absence of response to stimuli and the presence of rigor mortis and livor mortis forms a foundational framework within forensic science for approximating the time of death. This approach, blending the immediate with the gradual, allows forensic professionals to piece together the timeline of death's passage. It showcases the field's reliance on both observable physiological reactions and the natural processes that follow death, drawing on a blend of immediate assessments and the slow but steady markers that emerge in the hours and days post-mortem. Through this meticulous integration of signs, forensic science navigates the complexities of death, bridging the gap between the cessation of life and the unraveling of its final mysteries.

Conclusion – The narrative of Jesus' crucifixion, as captured within the Gospel accounts, highlights a moment of profound silence in response to external stimuli. This moment resonates deeply with the medical understanding of death's indicators. When viewed through the lens of modern medical science, this absence of reflexive reaction emerges as a pivotal marker in the determination of death. In contemporary clinical practice, this unyielding calmness, especially when it accompanies the disappearance of vital signs such as heartbeat and respiration, serves as a cornerstone in the framework for confirming death.

Today's medical field, equipped with sophisticated technologies, extends the capacity to diagnose death beyond mere observation, incorporating biochemical and electrical measurements that offer a precise window into the body's internal state. Techniques such as electroencephalography (EEG) and various forms of imaging provide detailed insights into the cessation of brain activity, while biochemical markers can indicate the breakdown of cellular processes.

Despite the advancements in medical technology that have expanded the toolkit for diagnosing death, the underlying principle that anchors these practices has not wavered. The absence of response to external stimuli is a testament to the cessation of the body's most fundamental life processes. This criterion, deeply rooted in the body's innate reflexive capabilities, continues to serve as a critical bridge between the observable external signs and the internal cessation of life functions.

Thus, the ancient observation documented in the Gospel narratives finds a parallel in the sophisticated practices of modern medicine, underscoring a continuity in humanity's approach to discerning the finality of life. It highlights how, despite the evolution of medical practices and technologies, the fundamental principles guiding the recognition of death remain steadfast, bridging centuries of medical thought and practice in understanding life's cessation.

Purpose of Breaking Legs in Crucifixion

The ancient practice of crurifragium, involving the deliberate fracturing of the legs of those undergoing crucifixion, adds a darkly profound layer to an execution method already steeped in brutality. Far from being an arbitrary act of violence, this procedure bore a calculated and ruthless intent: to accelerate the victim's demise through the mechanism of asphyxiation. The process of crucifixion itself, designed to maximize suffering and prolong the agony of death, becomes even more lethal with the infliction of such grievous injuries.[17]

In the grim dance of death that crucifixion represents, the victim's ability to push themselves up to breathe is crucial for prolonging life amidst unbearable pain. However, breaking the legs eradicates this last vestige of self-preservation. The victim plunges into an inexorable decline towards asphyxiation when they are deprived of the means to lift their body to draw in air. Each labored breath becomes a monumental struggle, as the body can no longer meet its demand for oxygen through the normal mechanics of breathing.

This act of breaking the legs, then, is not merely an addition to the torment but a methodically engineered step to ensure a quicker passage to death. It brings attention to the calculated cruelty that is embedded within the method of crucifixion, showcasing how human ingenuity has been utilized not solely for killing, but also for perfecting the process of dying into an art form of suffering. The physiological repercussions of such an action – coupled with the already excruciating conditions of crucifixion – paint a vivid picture of an execution method designed with the express purpose of inflicting maximum pain before yielding to the inevitability of death.

In the agonizing ordeal of crucifixion, the victims find themselves in a harrowing position, with their arms forcefully stretched and securely affixed either above their head or extended to the sides. This method of execution does not merely immobilize the victim. Still, it subjects their body to a relentless and excruciating strain, mainly targeting the chest muscles, the intercostal muscles nestled between the ribs, and the diaphragm. These muscles are pivotal for the act of breathing, a function that becomes a monumental challenge under these circumstances.

As the body dangles, suspended by the arms and shoulders, the weight of the victim's form becomes a cruel adversary. The mechanics of breathing in such a constrained posture demand an extraordinary effort. The crucified individual must engage in a desperate struggle against the very implements of their torment in order to draw a breath. They must pull upwards against the nails that pierce their limbs, an action that is a source of intense pain. Simultaneously, they must attempt to push their body upwards by exerting force on their legs, seeking to reduce the unbearable tension on their chest

---

[17] Ibid 5.

muscles and to allow for the expansion of their rib cage.[18]

This necessity to lift oneself, to make even the slightest movement to facilitate inhalation, involves not only the agonizing wounds inflicted by the nails but also battles against the overwhelming fatigue that engulfs the leg muscles. Each attempt to breathe becomes a feat of endurance, a painful exertion where the crucified must literally lift themselves against the pull of gravity and their bodily weight, all while trapped in the throes of relentless suffering and exhaustion. This cycle of pain and the struggle for each breath underscores the sheer brutality of crucifixion, marking it as one of history's most torturous methods of execution.

The relentless endeavor to sustain breath under the conditions of crucifixion initiates a vicious cycle of exhaustion. The constant, herculean effort required merely to draw air into the lungs subjects the victim's muscles to an unyielding state of tension and exertion. As time inexorably marches on, the muscles tasked with facilitating this basic, life-sustaining action begin to betray the individual, succumbing to fatigue with each passing moment.

This fatigue is not a simple weariness but a profound depletion of strength, as the muscles involved in breathing – the diaphragm, the intercostal muscles between the ribs, and even the muscles of the legs and chest – become increasingly unable to perform their functions. Each attempt to breathe grows incrementally more arduous than the preceding one as the constant struggle against gravity and the body's weight drains away the body's natural energy reserves.

Over time, the exhaustion accumulates and compounds, resulting in increasingly debilitating effects. In a dire predicament, the victim is faced with the challenge of mustering more effort with each breath. Amidst the cruel mechanics of crucifixion, the act of breathing, usually assumed effortless, turns into an arduous undertaking – an ongoing battle against the body's limitations. This cycle of exertion and exhaustion underlines the brutal efficiency of crucifixion as a method of execution, designed not just to kill but to torment.

The act of breaking the legs of a person subjected to crucifixion introduces a swift and grim progression toward death, primarily through the mechanism of asphyxiation. This brutal intervention strips the victim of their last recourse to mitigate the torturous effects of their position. Without using their legs, the individual is incapable of pushing upwards to relieve the relentless pressure exerted on their chest and diaphragm – muscles essential for breathing.

Before the breaking of the legs, the victim, despite enduring excruciating pain, could at least attempt to elevate their body slightly, affording themselves brief moments of respite wherein the ribcage could expand, and the lungs could fill with air. However, once deprived of this ability, the situation becomes dire. The pressure on the chest and diaphragm remains unrelieved, making the expansion of the lungs increasingly difficult with each attempted breath.

---

[18] Ibid 10.

This enforced immobility results in severe respiratory distress, as the victim struggles in vain for air that their body can no longer sufficiently draw. The diaphragm, now unable to contract and expand effectively, leaves the lungs gasping for breath. The rapid onset of asphyxia follows; it is a distressing condition where the body is deprived of oxygen, leading to a catastrophic failure of bodily functions.

In this state, the victim's plight escalates from a battle against pain and exhaustion to an inevitable suffocation. The breaking of the legs, therefore, not only exacerbates the physical suffering of the crucified but also hastens their death by directly inducing a fatal asphyxiation. This act, in its cruel efficiency, marks a dark and decisive step towards the end, ensuring that the victim's struggle for breath becomes a futile endeavor against the unyielding grip of death.

The grim aftermath of breaking the legs of a person crucified sets off a catastrophic cascade within their body, primarily characterized by hypoxia and hypercapnia – conditions that spell doom for the victim's physiological functions. The individual's respiratory system fails to adequately exchange air, depriving them of the ability to push upward for breath and leading to these critical imbalances in blood gases.

Hypoxia, the medical term for decreased oxygen levels in the blood, begins to take its toll on the body's tissues and organs, with the brain particularly vulnerable. Oxygen is the lifeblood of neural activity; without it, brain functions deteriorate. Initially, this may manifest as confusion and disorientation as the brain struggles to operate under oxygen-starved conditions. As hypoxia worsens, the victim may slip into unconsciousness, a dire sign that the brain can no longer sustain its normal operations.[19] If this condition persists without intervention, it inevitably leads to the cessation of heart function. The heart, starved of the oxygen necessary for its muscular contractions, simply cannot continue to pump, leading to cardiac arrest.

Simultaneously, the body faces the menace of hypercapnia – marked by increased carbon dioxide levels in the blood. Under normal circumstances, carbon dioxide is a waste product of the body's metabolism, efficiently expelled with each exhalation. However, with the crucified unable to breathe effectively, carbon dioxide accumulates, leading to respiratory acidosis. This condition disrupts the delicate acid-base balance of the body's internal environment, creating a state of increased acidity in the blood. Such an acidic milieu further destabilizes the body, exacerbating the already critical condition of the cardiovascular system. Hypoxia and acidosis exert additional pressure on the heart and blood vessels, pushing them closer to failure.

This lethal combination of hypoxia and hypercapnia, brought on by the inability to breathe properly after the breaking of the legs, thus accelerates the collapse of the body's vital systems. It is a slow, inexorable march toward death as the body's organs and systems fail one by one, overwhelmed by the lack of oxygen and the poisonous buildup of carbon dioxide. This process, stark and unforgiving, illustrates the brutal efficiency of crucifixion

---

[19] Ibid 12.

Chapter 8 – Confirmation of Jesus' Death

as a method of execution, designed not just to kill but to inflict maximum physiological torment along the way.

In the harrowing progression towards asphyxiation, the body's response is not merely a straightforward deprivation of air but unfolds as a multifaceted physiological crisis, culminating in cardiovascular collapse. Initially, the body instinctively launches into a desperate bid for survival, reacting to the plummeting oxygen levels by ramping up the heart rate and narrowing the peripheral blood vessels. This is a primal fight response aimed at preserving blood flow to vital organs and attempting to maximize the oxygen supply by redirecting it where it's most needed.

This compensatory mechanism, however, is only a temporary fix in the face of sustained oxygen deprivation. Given the circumstances, the heart, now beating at an accelerated pace, demands more oxygen – a demand that cannot be met. This vicious cycle puts the heart under immense strain as it works overtime in an environment where oxygen is critically scarce. Similarly, the constriction of peripheral blood vessels, while initially helpful in shunting blood to essential organs, ultimately increases blood pressure, further taxing an already overburdened cardiovascular system.

As these emergency measures become increasingly ineffective, the body's ability to cope with the oxygen shortfall begins to falter. The heart, strained beyond its capacity to compensate, starts to fail in its function of pumping blood efficiently throughout the body. This leads to a dramatic drop in blood pressure and a reduction in blood flow, signaling the onset of cardiovascular collapse. Blood, now poorly oxygenated and circulated, can no longer sustain the metabolic needs of the body's tissues and organs, precipitating a systemic failure.

The collapse of the cardiovascular system in the face of asphyxiation is a testament to the body's limits when confronted with severe oxygen deprivation. Despite the body's initial valiant efforts to adapt and survive, the relentless lack of oxygen creates an unsustainable situation, leading inevitably to a catastrophic breakdown of one of the body's most vital systems. This complex interplay of physiological reactions, from the heart's accelerated beating to the constriction of blood vessels, highlights the intricate and ultimately tragic nature of asphyxiation as a cause of death.

Asphyxiation, far from being a mere absence of air, triggers a catastrophic domino effect of physiological failures throughout the body, culminating in the breakdown of vital organs. The insidious duo of dwindling oxygen supplies and escalating carbon dioxide levels casts a wide net of destruction. Still, the brain, heart, and kidneys bear the brunt of this assault with particularly acute vulnerability.

The brain, the body's command center, is exquisitely sensitive to oxygen deprivation. Oxygen is the fuel that powers neuronal activity; without it, cognitive functions begin to sputter and fail. This lack of oxygen, known as hypoxia, sets off a cascade of cellular events that can lead to irreversible damage. Deprived of their essential sustenance, Neurons cannot maintain their normal operations, leading to a rapid decline in mental faculties and, ultimately, the cessation of all brain activity.

Simultaneously, the lack of a steady supply of oxygen throws the heart, which relies on it to pump blood effectively, into disarray. As oxygen levels plummet, the heart's muscle fibers struggle to contract and relax properly, undermining the heart's ability to circulate blood. This disruption in cardiac function further exacerbates the body's oxygen shortage, creating a vicious cycle of declining efficiency.

The kidneys, too, suffer under the conditions of asphyxiation. Tasked with filtering waste from the blood and maintaining the body's fluid and electrolyte balance, these organs depend on adequate blood flow and oxygenation. However, as the cardiovascular system falters and blood flow diminishes, the kidneys suffer from oxygen starvation and become overwhelmed by the accumulating toxins and acidosis. In this condition, the blood becomes too acidic. This toxic environment is antithetical to kidney function and can quickly lead to renal failure.

With prolonged exposure, the body's tissues and bloodstream being deprived of oxygen and inundated with carbon dioxide in this state of hypoxia and acidosis, multiple organ failure is inevitably precipitated. Each organ's struggle and subsequent failure further diminishes the body's overall function, leading to a systemic collapse. This tragic sequence underscores the profound and complex nature of asphyxiation, revealing it as a condition that mercilessly attacks the body's most critical systems, leaving devastation in its wake.

In the dire circumstances of asphyxiation, the body faces a critical disruption in its delicate biochemical balance, notably through the phenomenon of metabolic acidosis. This condition emerges not from a direct lack of oxygen but from the body's failure to rid itself of carbon dioxide, a waste product of its metabolic processes. Carbon dioxide accumulates in the bloodstream and transforms into carbonic acid, tipping the body's internal environment towards acidity.

This shift towards an acidic state, or acidosis, has profound implications for bodily function, particularly impacting the heart. The body's internal chemical balance finely tunes the heart's rhythm and ability to contract. Acidosis disrupts this balance, impairing the heart's electrical activity and weakening its contractions. This additional burden can be catastrophic in an already compromised system, where every heartbeat is crucial for survival. The heart, struggling to pump efficiently under the weight of acidosis, becomes even less capable of delivering the scant oxygen available to tissues and organs.

Moreover, metabolic acidosis sets off a chain reaction across multiple systems. In its attempt to compensate for the increased acidity, the body may deplete its reserves of essential minerals and buffers that help maintain pH balance. This depletion further destabilizes the internal environment, exacerbating the effects of oxygen deprivation and contributing to a faster decline in physiological functions.

The escalation of metabolic acidosis in the context of asphyxiation illustrates a grim turning point. As the body's mechanisms to expel carbon dioxide falter, the resultant acidosis not only directly undermines cardiac function but also accelerates the overall trajectory toward death. This intricate interplay between respiratory failure, carbon dioxide buildup, and acidosis highlights the complex and multi-faceted nature of asphyxiation, marking it as a condition that extends far beyond a mere inability to breathe, encompassing a broad spectrum of physiological failures that culminate in a tragic end.

Conclusion – The deliberate act of breaking the legs of an individual undergoing crucifixion emerges not merely as a brutal addition to an already torturous process but as a meticulously cruel method to expedite death through asphyxiation. A deeper exploration into the medical and physiological ramifications of this action sheds light on the severe impact it would have on the victim's capacity for respiration, setting off a rapid and torturous path toward death.

This practice, steeped in calculated brutality, significantly accelerates the onset of asphyxiation by robbing the victim of their last vestige of control over their breathing. With the legs broken, the crucified individual loses the ability to push upwards, a desperate measure that allows for momentary relief from the suffocating pressure on their chest and diaphragm. The absence of this ability plunges the victim into an inescapable predicament, where the body's demand for oxygen becomes futile against the relentless forces of gravity and physical exhaustion.

The ensuing physiological spiral – marked by a catastrophic failure to exchange air effectively, leading to hypoxia, hypercapnia, and metabolic acidosis – traps the victim in a web of acute distress and systemic collapse. Each breath, or the lack thereof, draws them closer to an inevitable demise characterized by multiple organ failure and cardiovascular collapse.

In this context, the method of breaking legs stands as a stark testament to the cruelty embedded in the practice of crucifixion. The technique aimed not only to end life but also to magnify suffering, ensuring that the journey towards death was filled with unparalleled agony. This understanding underscores the grim efficiency of crucifixion as a means of execution, highlighting its capacity to inflict maximum torment before yielding the release of death. Through this lens, the act transcends its historical context, serving as a poignant reminder of the depths of human cruelty and the profound struggle for breath that marked the final moments of those subjected to such a fate.

CHAPTER 9

LIMITATION IN ANALYZING THE DATA

Before coming to a conclusion about what happened to Jesus, it is important to address some limitations in the analysis presented thus far. We must remain objective about the reliability of the texts that describe the events leading up to Jesus' death, as this is crucial in developing possible diagnoses that could explain his passing. It's important to remember that the purpose of this medical inquiry is to investigate the possibility that Jesus died from natural causes, which is essential in considering his resurrection. Without a physical body, there can be no resurrection. Therefore, we speculate and analyze the potential outcomes of the trauma reported in the ancient texts known as the Gospels.

Medicine in Antiquity

Jesus was under Roman jurisdiction when he experienced all the trauma that led to his death. We are basing all the medical diagnoses on the Roman knowledge of medicine, where Roman soldiers trained in the art of torment, applied all kinds of torture methods that led to his death.

The knowledge base of ancient Roman medicine, heavily reliant on the foundational works of Greek physicians such as Hippocrates (circa 460–370 BCE) and Galen (circa 129–216 CE), was underpinned by principles and theories that shaped medical practice for centuries. Through observation, documentation, and speculation, these practitioners laid the groundwork for medical diagnosis and treatment in their time. Their texts and theories extended well into the Roman era, where people revered, studied, and expanded upon them. However, contemporary medical understanding and the absence of tools and technologies significantly restricted the diagnostic capabilities of the time, which modern medicine takes for granted.

Foundational works of Hippocrates that we have today is a collection of around 70 early medical works from ancient Greece, which are attributed to

Hippocrates and his followers, known as Hippocratic Corpus.[1] These texts not only cover a wide range of medical topics, including diagnosis, epidemiology, pediatrics, and surgery but also articulate the Hippocratic Oath. The Hippocratic approach emphasized careful observation and detailed note-taking, asserting that diseases had natural, not supernatural, causes.

Galen, Greek physician, writer, and philosopher who exercised a dominant influence on medical theory and practice in Europe, developed the theory of the four humors. He conducted numerous dissections (though mainly on animals due to Roman laws against human dissection) and experiments to explore anatomy and physiology.[2] Galen's extensive writings included detailed descriptions of the humoral theory, asserting that a balance among the four humors was essential for health. His works remained a dominant influence in medicine well into the Renaissance.

The "Theory of the Four Humors" governed human health and temperament based on the balance of four bodily fluids: blood, phlegm, yellow bile, and black bile. According to the "Theory of the Four Humors," people believed that the balance of four bodily fluids (blood, phlegm, yellow bile, and black bile) governed human health and temperament. They associated each humor with specific qualities (hot, cold, wet, and dry) and elements (air, water, fire, and earth), and they believed that imbalances among these humors caused disease. For example:

- It was believed that an excess of blood, associated with spring and air, caused a sanguine temperament, characterized by optimism and sociability, but imbalance could lead to fevers and inflammatory conditions.
- Phlegm, linked to winter and water, was associated with a phlegmatic temperament, seen as calm and unemotional, but excess could result in colds, flu, and other phlegmatic conditions.
- The belief was that yellow bile, corresponding to summer and fire, would lead to a choleric temperament marked by ambition and irritability, with imbalances causing digestive issues and anger.
- It was believed that black bile, associated with autumn and earth, resulted in a melancholic temperament, which was characterized by introspection and sadness. Imbalances of black bile would now be recognized as depressive symptoms.

While the humoral theory and the reliance on observational techniques limited the accuracy of ancient diagnoses by modern standards, these approaches represented significant advancements in medical thinking.[3] They moved away from supernatural explanations for disease and laid the groundwork for systematic observation and the classification of illnesses.

---

[1] Jouanna, J. (1999). *Hippocrates*. Johns Hopkins University Press.
[2] Nutton, V. (2013). *Ancient Medicine*. Routledge.
[3] Pormann, P.E., & Savage-Smith, E. (2007). *Medieval Islamic Medicine*.

Treatments were based on restoring humor balance through diet, exercise, bloodletting, and the use of herbal remedies.

The absence of modern diagnostic tools, such as imaging technologies and laboratory tests, meant that ancient physicians had to rely heavily on their senses and the patient's description of symptoms. This led to a holistic approach to patient care, considering not just physical symptoms but also the emotional and environmental context of the patient's life.

In brief, the knowledge base of ancient Roman medicine, deeply influenced by Greek predecessors, was a blend of empirical observation, philosophical interpretations of health and disease, and practical treatments aimed at balancing bodily humors. Despite its limitations, this foundation contributed significantly to the evolution of medical science, emphasizing natural causes for diseases and the importance of careful observation and documentation in medical practice.

The diagnostic methods that Roman physicians employed, while primitive by modern standards, were advanced for their time and laid the groundwork for future medical practices. These methods were primarily based on direct observation, patient interviews, and a rudimentary understanding of anatomy and physiology. Despite the lack of sophisticated diagnostic tools and a comprehensive understanding of disease mechanisms, Roman doctors developed a systematic approach to diagnosing and treating illnesses.

Observation was a critical component of the diagnostic process. Physicians relied heavily on visual inspection of the patient's body, noting the color, texture, and appearance of the skin, eyes, and other accessible parts. They observed physical signs such as swelling, rashes, or abnormal discharges, which provided clues to the underlying condition.[4]

Patient interviews complemented these observations, allowing physicians to gather information on symptoms, pain, dietary habits, lifestyle, and environmental factors that could influence health. These conversations helped build a holistic picture of the patient's condition, incorporating both the physical and social aspects of their lives.

The Roman medical community possessed limited knowledge of anatomy and physiology compared to today's understanding, but it was still significant for its time. This understanding was primarily based on the writings of earlier Greek physicians like Hippocrates and Galen, who, despite the limitations imposed by societal norms against dissection, made considerable strides in mapping the human body and theorizing about its functions. Galen, in particular, conducted animal dissections that he extrapolated to humans, offering detailed descriptions of the cardiovascular system, nerves, and muscles. However, inaccuracies persisted due to the lack of human dissection.

---

[4] Jackson, R. (1988). *Doctors and Diseases in the Roman Empire*. University of Oklahoma Press.

Chapter 9 – Limitation in Analyzing the Data

Diagnostic Techniques

Roman physicians employed various techniques to aid their diagnoses[5]:

- **Pulse Diagnosis**: Although rudimentary, the examination of the pulse was a key diagnostic tool, used to assess the heart's condition and, by extension, the patient's overall vitality.
- **Urine Examination**: The analysis of urine was another common practice. Physicians observed the urine's color, consistency, sediment, and sometimes even taste, to diagnose conditions, especially those related to the kidneys and bladder.
- **Palpation and Percussion**: Touching and tapping different body parts allowed physicians to identify areas of tenderness, swelling, or abnormality, aiding in the diagnosis of internal diseases.
- **Diet and Lifestyle Analysis**: Recognizing the impact of diet and lifestyle on health, Roman doctors often inquired about these aspects, understanding that changes in these areas could both cause and remedy ill health.

Roman physicians' ability to diagnose diseases was inherently limited due to the absence of modern diagnostic tools. Due to the absence of modern diagnostic tools, Roman physicians had an inherently limited ability to diagnose diseases as they lacked knowledge of the germ theory of disease, which would not be developed until the 19th century, and had no access to microscopes, blood tests, imaging technologies, or other tools that form the cornerstone of modern diagnostics. This lack of understanding about infectious agents and the internal workings of the body often led to diagnoses based on symptoms rather than underlying causes.

Despite these limitations, Roman physicians made considerable advances in medical practice. Their emphasis on observation, detailed patient histories, and the holistic approach to diagnosis and treatment laid the foundations for future developments in medical science. The systematic approaches to health and disease they developed, based on the tools and knowledge available, contributed significantly to the care of patients in their time and influenced medical practice for centuries to come.

The accuracy and interpretation of medical diagnoses from ancient Roman times are significantly influenced by the state of preservation, quality of ancient records, and the subsequent handling of these texts through centuries. The process of record-keeping in Roman times, coupled with the challenges faced in modern interpretation, presents a complex picture that requires careful consideration.

To summarize, ancient Roman medical diagnoses were based on the best knowledge and techniques available at the time, but they cannot be directly compared to modern standards of accuracy. The ancient Romans made significant observations about health and disease, but the limitations

---

[5] King, H. (2001). *Greek and Roman Medicine*. Bristol Classical Press.

of their medical knowledge and diagnostic capabilities mean that their accuracy would vary greatly by today's criteria. Modern analysis of ancient texts and archaeological findings can provide insights into these diagnoses, but researchers must interpret them within the historical and cultural context of the time.

Our aim was to highlight the fact that the responsibilities of Roman soldiers were not as basic as some people may assume, especially when it came to tortures. In ancient times, people viewed medicine as a form of science, and doctors took an active role in researching and developing diagnostic and treatment methods. The methods used by Roman soldiers to determine Jesus' death were quite advanced, contrary to popular belief. They had established landmarks and techniques to determine whether someone was deceased.

Can You Diagnose it Based on Descriptive Facts?

Medical diagnosis based on descriptive facts, such as symptoms, patient history, and physical examinations, can be quite accurate but also depends on several factors. A medical diagnosis's accuracy relies on the experience and expertise of the healthcare provider, the specificity and clarity of the symptoms, the presence of common or rare diseases, and the use of additional diagnostic tools (like lab tests and imaging studies) to confirm the initial assessment.

Generally speaking, an experienced clinicians can often make accurate diagnoses with minimal information based on pattern recognition from past encounters with similar cases. However, even experienced providers can misdiagnose conditions that present atypically. Some conditions have very specific symptoms that can lead to a straightforward diagnosis. However, many conditions share common symptoms (like fever, fatigue, or pain), making it harder to diagnose based solely on descriptive facts without further testing.

A thorough patient history, including past medical history, medications, lifestyle, and family history, can significantly influence the accuracy of a diagnosis. Certain diseases have genetic components or are more common in specific lifestyles or exposures. Generally speaking, while the initial medical diagnosis is based on descriptive facts, especially for common conditions, it is crucial to use a combination of patient history, physical examination findings, and additional diagnostic tests to ensure the most accurate diagnosis.

Many medical providers will acknowledge that patient history and presentation play a significant role in making 70% of the diagnosis. The physical exam will confirm or put the provider in the new line of thought in the development of a diagnosis. In Jesus case, just the history and presentation as recorded in the Gospels, gave at least 70% probability of the diagnosis developed in the context of his death.

Limitation of the Analysis

The endeavor to apply modern medical insights to the interpretation of ancient texts is fraught with inherent challenges. The authors of these texts did not write them with the precision or intent of modern clinical documentation, even though they contain descriptions of physical symptoms, injuries, or phenomena. This mismatch between the ancient narrative style and contemporary medical analysis introduces several limitations that significantly affect the reliability and specificity of such interpretations. Here, we explore these limitations in detail, focusing on the medical aspects.

Speculative Nature of Interpretations

The speculative nature of interpreting ancient texts, especially those concerning health conditions, injuries, or physical phenomena, is a significant challenge for historians, medical professionals, and scholars. Historians, medical professionals, and scholars find it challenging to interpret ancient texts, especially those related to health conditions, injuries, or physical phenomena, primarily due to the frequent use of metaphorical, symbolic, or simplified language by ancient authors to describe medical conditions. This language can be quite different from the detailed, clinical observations that are typical of modern medicine. This divergence in descriptive approach leads to a range of interpretative challenges:

Metaphorical and symbolic language – Ancient texts often use metaphorical or symbolic language to describe illness or physical conditions, which can obscure the actual medical realities they intend to convey. For example, a text might describe someone as being "consumed by fever" or "overcome by a chill," phrases that convey a general idea of illness but lack specific diagnostic details. Such descriptions can lead to interpretations that are heavily influenced by the reader's own cultural, medical, and historical knowledge, potentially diverging significantly from the original author's intent.[6]

Simplification and generalization – In many cases, ancient descriptions of medical conditions are highly simplified or generalized. This lack of specificity can make it difficult to identify the exact nature of the condition being described. For instance, ancient texts might refer to "leprosy" for any number of skin diseases, not just the condition known as leprosy today. Similarly, "consumption" was a term broadly applied to conditions leading to wasting and weight loss, which could encompass everything from tuberculosis to cancer, without the detailed differentiation found in modern

---

[6] Asclepius, L. & Hippocrates, G. (2018). *Metaphors of Medicine: Interpreting Disease in Ancient Rome.* Cambridge University Press.

diagnostics.[7]

Multiple interpretations – The combination of metaphorical language and lack of specific detail means that the same ancient description can lead to multiple, sometimes conflicting, interpretations. Different scholars might read the same text and conclude that it describes entirely different conditions based on their own expertise, the context in which they interpret the text, or the assumptions they make about the author's understanding of medicine.

Challenges in translation – The translation of ancient texts presents another layer of complexity. The original language might contain terms or concepts that have no direct equivalent in modern languages, leading translators to choose words that approximate the original meaning. These choices can significantly affect how modern readers interpret descriptions of illness or injury. Additionally, translations can reflect the translator's own biases or interpretations, further complicating the understanding of the original text.

The role of cultural context – The cultural context in which these texts were written also crucially influences their interpretation. What one culture might consider a sign of serious illness, another might see as a minor ailment or even a spiritual or psychological condition. Understanding the cultural beliefs, medical knowledge, and societal norms of the time is essential for accurately interpreting descriptions of health conditions in ancient texts.[8]

Cultural and Historical Context

When interpreting ancient texts, especially descriptions of disease, injury, and physiology, the speculative nature is significantly influenced by the cultural and historical context of the period in which these texts were written. Ancient civilizations had their own unique understandings of the human body, health, and illness, which were shaped by their beliefs, knowledge, and the technological limitations of their time. This divergence from modern medical understandings presents several challenges and considerations:

Ancient understandings of disease and physiology – In ancient times, individuals commonly believed that diseases were understood by intertwining physical symptoms with spiritual, astrological, or humoral theories. For instance, the Greeks and Romans attributed disease to an imbalance among blood, phlegm, black bile, and yellow bile, according to their theory of the four humors. Similarly, many cultures attributed certain illnesses to divine punishment or the influence of evil spirits. The interpretations of diseases and their treatments in ancient texts were deeply influenced by the philosophical and religious beliefs of the time, which were

---

[7] Thompson, C. & Venables, R. (2017). *Cultural Contexts of Health: The Use of Narrative in Medicine*. Brill.
[8] Markham, J. (2016). *Illness in Antiquity: A Cultural and Linguistic History*. Ashgate.

rooted in these beliefs.[9]

Cultural interpretations of injury and healing – The treatment and understanding of injuries varied significantly across different cultures and historical periods. Treatments could range from the application of herbs and surgeries to rituals intended to appease gods or spirits. For example, different cultures performed trepanation, the practice of drilling holes into the skull to treat various conditions, with varying justifications, such as releasing evil spirits or treating physical trauma. These practices reflect not only the medical knowledge of the time but also the cultural attitudes towards injury, healing, and the human body.[10]

The impact of historical context on medical descriptions – The historical context, including the available technology, societal structure, and prevailing philosophical or religious doctrines, played a crucial role in shaping medical knowledge and its documentation. Those with the best theoretical or practical understanding of medicine often wrote ancient medical texts within their culture, such as priests, philosophers, or physicians. However, their ability to observe, describe, and the lack of advanced tools limited treat diseases like microscopes or diagnostic machines that are taken for granted in modern medicine. This limitation means that ancient descriptions of medical conditions are often vague, symptom-based, and open to interpretation.[11]

Challenges in interpreting ancient medical texts – Given these vast differences in understanding, interpreting ancient medical texts requires a careful consideration of the cultural and historical context. Modern scholars must navigate these texts, recognizing that ancient authors were operating within their own frameworks of knowledge and belief systems. In order to understand the ancient texts within their own frameworks of knowledge and belief systems, scholars often rely on speculative interpretation to translate ancient descriptions of symptoms, treatments, and outcomes into terms that can be understood within the context of contemporary medical science. However, this process is inherently speculative and can lead to multiple, sometimes conflicting, interpretations of the same texts.[12]

Challenges in Applying Modern Medical Knowledge

When examining historical descriptions of disease, injury, or medical practices, scholars and medical historians face the challenge of anachronistic bias. In examining historical descriptions of disease, injury, or

---

[9] Scarborough, J. (1992). *Medical and Biological Terminologies: Classical Origins.* University of Oklahoma Press.
[10] King, H. (1998). *Health in Antiquity.* Routledge.
[11] Porter, R. (1997). *The Greatest Benefit to Mankind: A Medical History of Humanity.* W.W. Norton & Company.
[12] Bynum, W.F., and Porter, R. (Eds.). (1993). *Companion Encyclopedia of the History of Medicine.* Routledge.

medical practices, scholars and medical historians encounter the challenge of anachronistic bias. This arises when they incorrectly apply modern medical concepts, understandings, or terminologies to interpret ancient texts or practices, resulting in assumptions about the level of anatomical or physiological knowledge that historical figures could not have possessed. This bias can distort our understanding of ancient medical practices and theories, presenting several specific challenges:

Misinterpretation of ancient medical terms – Anachronistic bias can lead to the misinterpretation of ancient medical terms, where individuals impose modern meanings on historical words that likely had different implications. For example, ancient medical texts might use a term that superficially resembles a modern medical condition but, in the context of the time, referred to a symptom complex or a set of conditions not aligned with contemporary understanding.[13]

Overestimation of historical medical knowledge – Applying modern medical knowledge to ancient descriptions can result in an overestimation of the historical understanding of anatomy and physiology. While ancient physicians made significant observations and developed complex theories, their insights were limited by the observational technologies available and the philosophical and religious frameworks that shaped scientific inquiry. Modern analyses that fail to account for these limitations may attribute a level of precision or insight to ancient practitioners that was not achievable at the time.[14]

Misunderstanding of historical practices – Anachronistic interpretations can also lead to misunderstandings of historical medical practices. Evaluating procedures or treatments described in ancient texts against modern standards of efficacy and ethics, without considering the theoretical rationales or the empirical observations that underpinned them, might lead to misunderstandings. This can result in a skewed perception of these practices, either undervaluing their effectiveness within their historical context or unfairly criticizing them for not aligning with contemporary medical principles.[15]

Projection of modern diagnoses – There is a tendency to diagnose historical figures or conditions described in ancient texts with modern diseases, ignoring the fact that disease presentations can change over time due to environmental, genetic, and social factors. By projecting modern diagnoses onto historical descriptions, it fails to consider the historical evolution of illnesses and how they were understood and categorized, thus assuming a continuity of disease that may not exist.[16]

---

[13] King, H. (1995). *Health in Antiquity*. Routledge.
[14] Horden, P. (2000). *The History of Medicine: A Beginner's Guide*. Oxford University Press.
[15] Ibid 14.
[16] Ibid 12.

Solutions and considerations – To mitigate the challenges posed by anachronistic bias, scholars must adopt a multidisciplinary approach that incorporates historical, cultural, and medical perspectives. This includes:

- A careful analysis of the language and terminology used in ancient texts, with attention to their historical meanings and contexts.
- An appreciation for the limitations of historical medical knowledge and an understanding of the theoretical frameworks within which ancient practitioners operated.
- A critical examination of historical medical practices and treatments, acknowledging their empirical bases and the contexts that gave rise to them.
- A nuanced approach to diagnosing historical conditions, recognizing the potential for disease presentations and understandings to change over time.

By acknowledging and addressing anachronistic bias, researchers can develop a more accurate and respectful understanding of historical medical practices, avoiding the imposition of modern concepts and categories on ancient descriptions and theories. This approach not only enriches our comprehension of the history of medicine but also highlights the evolution of medical knowledge and its contextual dependencies.

Diagnostic Oversimplification

The application of modern medical knowledge to interpret ancient descriptions of disease and treatment presents a significant challenge, particularly in the context of diagnostic oversimplification. Modern diagnostics are the product of centuries of medical advancement, relying heavily on a multidisciplinary approach that includes clinical examination, laboratory tests, radiological imaging, and more recently, genetic testing. This comprehensive approach allows for precise identification and understanding of diseases, a luxury not available in ancient times. The attempt to apply these standards to ancient descriptions without the aid of such tools retroactively can lead to oversimplified or inaccurate medical conjectures for several reasons.

Lack of diagnostic tools – Ancient medical practitioners lacked diagnostic tools and relied on observations of visible symptoms and patient reports for their diagnoses. Without access to laboratory tests, they could not confirm the presence of pathogens, metabolic imbalances, or genetic conditions. Similarly, they often inferred internal injuries, organ pathologies, and other conditions instead of directly observing them, due to the absence of imaging studies. This limitation necessitated a reliance on symptomatic descriptions that may appear vague or oversimplified by modern standards.[17]

---

[17] Ibid 2.

Different disease frameworks – Ancient times had fundamentally different conceptual frameworks for understanding diseases compared to modern medicine. People in ancient times used humoral theory or spiritual beliefs to categorize and explain diseases, instead of relying on the germ theory or cellular pathology that underpins modern diagnostics. As a result, ancient descriptions of illness often reflect these theoretical understandings, making direct comparison to modern disease categories challenging and potentially misleading.[18]

Evolution of diseases – The nature and presentation of diseases can evolve over time due to changes in environment, lifestyle, and human genetics. Diseases common in ancient times may present differently in the modern world, or they may have become rare or extinct. Similarly, new diseases have emerged that were not present in the ancient world. Attempting to match ancient descriptions of diseases with modern conditions can overlook these historical changes in disease epidemiology and presentation.[19]

Interpretative Bias – Interpreting ancient medical texts with a modern medical mindset can introduce bias, leading to the assumption that ancient practitioners were describing diseases in ways that align with contemporary understandings. This can result in the projection of modern disease categories onto ancient descriptions, ignoring the historical and cultural context that shaped these descriptions. Such bias can oversimplify complex conditions or misinterpret the nature of ancient medical knowledge.[20]

Solutions and Considerations – To address the challenge of diagnostic oversimplification, scholars and medical historians must adopt a nuanced approach that respects the historical context and recognizes the limitations of both ancient and modern medical knowledge. This includes:

- **Critical analysis of ancient texts:** When critically analyzing ancient texts, interpreters should carefully consider the language, metaphors, and theoretical frameworks used in ancient medical descriptions. They should seek to understand how diseases were conceptualized within their historical context.
- **Interdisciplinary collaboration:** Combining expertise from medical historians, philologists, archaeologists, and modern clinicians can provide a more comprehensive understanding of ancient medical practices and descriptions.
- **Awareness of historical disease evolution:** Recognizing that diseases can change over time allows for a more flexible approach to

---

[18] Ibid 6.
[19] Li, J. (2019). *Evolution of Diseases in Historical Perspective*. Springer.
[20] Siraisi, N.G. (1990). *Medieval and Early Renaissance Medicine: An Introduction to Knowledge and Practice*. University of Chicago Press.

matching ancient descriptions with modern conditions, acknowledging that direct correlations may not always be possible or appropriate.
- **Cultural sensitivity:** can help us understand that medical practices and disease interpretations are deeply embedded in cultural beliefs and practices, thus avoiding anachronistic judgments and appreciating the historical value of ancient medical knowledge.

By acknowledging and addressing these challenges, researchers can better navigate the complexities of applying modern medical knowledge to ancient descriptions, fostering a more accurate and respectful understanding of historical medical practices.

Variability in Individual Responses

The concept of variability in individual responses to trauma, disease, and environmental stresses is fundamental to understanding human health and pathology. A combination of genetic, environmental, physiological, and anatomical factors influences this variability. Here, we'll dive deeper into the nuances of physiological and anatomical variations, highlighting why ancient accounts often struggle to fully capture this complexity.

Physiological variability – Physiological variability refers to the differences in how individuals' bodies function and respond to external and internal stimuli. This encompasses a wide range of processes, including metabolic rates, immune system effectiveness, hormonal balances, and the capacity for tissue repair and regeneration.[21] Factors contributing to physiological variability include:

1. **Genetic Differences**: Genetic makeup plays a crucial role in determining an individual's susceptibility to diseases, response to medications, and recovery from injuries. For instance, variations in the CYP450 enzymes affect drug metabolism, influencing drug efficacy and risk of side effects.[22]
2. **Age and Developmental Stage**: Age significantly affects physiological responses. For example, young children and the elderly often have weaker immune responses, making them more susceptible to infections and diseases.[23]
3. **Lifestyle and Environmental Exposures**: Diet, exercise, exposure to pollutants, and stress levels can all influence physiological responses. For example, chronic stress can suppress the immune system, while regular physical activity can enhance it.

---

[21] Mora, S., et al. "*Physical activity and reduced risk of cardiovascular diseases: biological and epidemiological evidence.*" The Lancet 366.9499 (2005): 1789-1799.
[22] Zhou, S. F., et al. "*Polymorphism of human cytochrome P450 enzymes and its clinical impact.*" Drug Metabolism Reviews 41.2 (2009): 89-295.
[23] Nikolich-Žugich, J., et al. "*Ageing and life-long maintenance of T-cell subsets in the face of latent persistent infections.*" Nature Reviews Immunology 8.7 (2008): 512-522.

4. **Pre-existing Health Conditions**: Individuals with pre-existing health conditions may respond differently to new stresses or diseases due to compromised organ systems or ongoing treatments.

Anatomical variability – Anatomical variability refers to the differences in structure and organization of the body and its organs. These differences can affect how individuals experience and recover from trauma or disease.[24] Key aspects include:

1. **Body Composition**: Variations in muscle mass, fat distribution, and bone density can influence susceptibility to injury, disease outcomes, and medication dosages.
2. **Organ Size and Function**: Differences in the size and efficiency of organs like the liver and kidneys can affect drug metabolism and the body's ability to detoxify substances.
3. **Variations in Blood Vessels and Nerve Pathways**: Anatomical variations in the circulatory and nervous systems can affect the spread of diseases and the experience of pain.

Challenges of ancient accounts – Ancient medical accounts often lack the detail and scientific understanding to fully appreciate the complexity of individual variability. These accounts were based on observations and theories that did not have access to modern diagnostic tools or a genetic understanding of disease. Consequently, ancient treatments did not tailor to individual differences and instead typically generalized. For example, the humoral theory of medicine, prevalent in ancient Greek and medieval medicine, attributed disease to imbalances in four bodily fluids and did not account for genetic or anatomical variations.

Moreover, the absence of detailed case studies and the limited understanding of anatomy and physiology meant that ancient physicians could not account for the intricate interplay of factors that influence individual responses to disease and treatment. As a result, while ancient texts provide valuable insights into the medical knowledge and practices of the past, they fall short in capturing the full spectrum of human biological variability.

In conclusion, the variability in individual responses to trauma, disease, and environmental stresses underscores the importance of personalized medicine. Understanding and accounting for physiological and anatomical variations are crucial for diagnosing conditions accurately, predicting outcomes, and tailoring treatments to individual needs – a concept that is increasingly feasible with advances in genomics, biotechnology, and data analytics, but was largely beyond the reach of ancient medical practice.

Lack of Detailed Case Histories

The variability in individual responses to diseases, treatments, and environmental factors is a cornerstone of contemporary medical

---

[24] Bergman, R. A., et al. *Variants in Human Anatomy*. 13th ed., e-Anatomy, 2016.

Chapter 9 – Limitation in Analyzing the Data

understanding. A crucial aspect that underscores this variability is the detailed documentation of case histories, which include symptoms, progression of conditions, treatment responses, and outcomes. However, ancient medical texts often lack these comprehensive case histories, presenting significant challenges in fully understanding and interpreting historical accounts of medical conditions and practices.[25]

Importance of detailed case histories – Detailed case histories are vital for several reasons:

1. **Diagnostic Insight**: They provide in-depth insights into the symptoms and progression of diseases, helping in the formulation of accurate diagnoses.
2. **Treatment Evaluation**: Documenting responses to treatments over time allows for the assessment of their efficacy and the identification of side effects or complications.
3. **Personalized Medicine**: Recognizing the variability in individual responses helps tailor treatments to individual needs, enhancing their effectiveness and minimizing risks.
4. **Educational Value**: Case histories serve as educational tools, illustrating the complexities of disease management and the importance of considering individual patient factors.
5. **Research and Development**: They contribute to medical research by offering data that can be analyzed for patterns, leading to new medical insights and the development of new treatments

Challenges posed by the lack of detailed case histories in ancient texts – Ancient medical texts, while invaluable for their historical and cultural insights, often provide only cursory descriptions of medical conditions, with little to no detail on the patient's history, symptom progression, or response to treatment.[26] This absence poses several challenges:

1. **Limited Understanding of Conditions**: Without detailed case histories, it's difficult to ascertain the exact nature of the conditions described, their severity, and their progression over time.
2. **Generalization vs. Individuality**: Ancient treatments were often based on generalized theories of disease (such as the humoral theory in Greek medicine) rather than individual patient conditions and responses, which may lead to misconceptions about the effectiveness or applicability of these treatments.
3. **Interpretation and Translation Issues**: The lack of detail increases the risk of misinterpretation of ancient texts, especially when

---

[25] Siraisi, Nancy G. *Medieval and Early Renaissance Medicine: An Introduction to Knowledge and Practice*. University of Chicago Press, 1990.
[26] Hunter, Kathryn Montgomery. *Doctors' Stories: The Narrative Structure of Medical Knowledge*. Princeton University Press, 1991.

translations are involved, as subtle nuances about conditions or treatments may be lost.[27]
4. **Comparative Historical Medical Research**: For researchers attempting to understand the evolution of medical knowledge and practices, the absence of detailed case histories makes it challenging to compare ancient practices with modern understandings of medicine directly.

Contemporary relevance – Today, the detailed documentation of case histories is a fundamental aspect of medical practice and research. It allows for a nuanced understanding of diseases and treatments, facilitating advancements in medical science and the customization of patient care. The contrast with ancient medical documentation highlights not only the evolution of medical knowledge and practices but also underscores the importance of individual patient data in advancing healthcare.

In conclusion, while ancient medical texts provide a window into the past, their lack of detailed case histories limits our ability to fully understand and appreciate the complexities of historical medical practices and the conditions they aimed to treat. This gap underscores the evolution of medical documentation and the increasingly personalized approach to medicine that modern case histories facilitate.

Ethical and Methodological Concerns

The interpretation of ancient medical texts through a contemporary lens raises significant ethical and methodological concerns, particularly regarding the projection of modern values and ethical standards onto historical contexts. This issue is complex and multifaceted, involving the risk of misinterpreting or distorting ancient practices, beliefs, and descriptions based on current medical knowledge, ethics, and societal norms. Understanding these concerns requires a closer examination of the nuances involved in historical medical interpretation.[28]

Ethical Concerns

1. **Anachronism**: Applying modern ethical standards to ancient contexts can be anachronistic, meaning it imposes present-day values on periods where these concepts did not exist or were fundamentally different. For example, concepts of patient consent, confidentiality, or the Hippocratic Oath's interpretation have evolved significantly over millennia. Misapplying contemporary ethics can lead to misunderstandings about the intentions and practices of ancient healers.[29]

---

[27] Hamburg, Margaret A., and Francis S. Collins. "*The Path to Personalized Medicine*." New England Journal of Medicine 363, no. 4 (2010): 301-304.
[28] Jonsen, Albert R. *The Birth of Bioethics*. Oxford University Press, 1998.
[29] Nutton, Vivian. *Ancient Medicine*. Routledge, 2012.

2. **Cultural Relativism**: Ancient texts are products of their time, reflecting the cultural, social, and philosophical norms of their societies. Modern interpreters must be cautious not to judge these texts solely by today's ethical standards, which might be incongruent with the historical context. Recognizing the cultural relativism inherent in these texts is crucial to avoid ethnocentric biases that can skew understanding and appreciation of ancient medical practices.

Methodological Concerns

1. **Historical Contextualization**: Proper interpretation of ancient medical texts requires a deep understanding of the historical context, including the prevailing medical theories, societal structures, and philosophical beliefs of the time. Without this contextualization, there's a risk of projecting modern medical paradigms onto the ancient texts, potentially misinterpreting the original intentions and applications of ancient practices.[30]
2. **Translation and Interpretation**: Translators across centuries have translated many ancient texts multiple times, potentially introducing biases or reinterpretations based on their understanding and contemporary values of their period. Modern readers must navigate these layers of interpretation, which can obscure the original text's meaning.
3. **Evolution of Medical Language**: The meanings of medical terms and concepts have evolved, with some ancient terms having no direct modern equivalent or significantly different implications. Interpreters must carefully navigate the linguistic shifts and the evolution of medical language in order to understand ancient texts accurately.[31]
4. **Research Bias**: Researchers may enter the study of ancient texts with preconceived notions shaped by modern medical and ethical standards. This bias can influence researchers, causing them to emphasize certain aspects of the texts or overlook others. Consequently, their interpretations may be skewed, highlighting congruence with modern practices while neglecting fundamental differences.

Addressing the Concerns

To mitigate these ethical and methodological concerns, scholars and interpreters of ancient medical texts should adopt a multidisciplinary approach, incorporating historical, cultural, linguistic, and medical expertise.[32] This approach should aim to:

---

[30] Lloyd, G.E.R. *Methods and Problems in Greek Science: Selected Papers*. Cambridge University Press, 1991.
[31] Harris, Stephen L. *Understanding the Bible*. McGraw-Hill Education, 2010.
[32] King, Helen. *The Disease of Virgins: Green Sickness, Chlorosis and the Problems of Puberty*. Routledge, 2004.

- Acknowledge and critically examine one's biases and the potential for anachronism.
- Engage with the texts in their original languages, where possible, to minimize translation biases.
- Collaborate with historians, philologists, and anthropologists to enrich the medical interpretation with broader cultural and historical insights.
- Promote transparency in the interpretative process, explicitly acknowledging the interpretive choices and the rationale behind them.

In conclusion, interpreting ancient medical texts through a modern lens requires careful consideration of ethical and methodological concerns to avoid projecting contemporary values onto historical contexts. By acknowledging these complexities and striving for a nuanced, interdisciplinary approach, scholars can more faithfully represent and understand the medical practices and beliefs of the past.

Interdisciplinary Limitations

The integration of interdisciplinary approaches to analyze ancient medical texts is a methodological advancement that brings together expertise from various fields such as medicine, history, linguistics, and anthropology. This collaborative effort is designed to provide a more nuanced and comprehensive understanding of ancient medical practices, beliefs, and descriptions. However, this approach also unveils significant ethical and methodological concerns, particularly regarding the limitations inherent within each contributing discipline. Understanding these limitations is crucial for accurately interpreting ancient texts and for the ethical representation of historical medical knowledge.[33]

Limitations of medical professionals – Medical professionals bring invaluable insights into the biological and pathological aspects of descriptions found in ancient texts. Their expertise is critical for interpreting the medical implications of various conditions described historically. However, they may face challenges including:

1. **Historical Context**: Medical professionals might not have a full understanding of the historical context in which these texts were written, including the societal norms, medical theories, and philosophies of the time. Without this context, there's a risk of misinterpreting the intention behind medical practices or the significance of certain descriptions.[34]

2. **Linguistic Nuances**: Ancient languages can contain terms and concepts that do not have direct equivalents in modern languages.

---

[33] Jones, David S., Jeremy A. Greene, and Scott H. Podolsky. "*Making the Case for History in Medical Education.*" Journal of the History of Medicine and Allied Sciences 70, no. 4 (2015): 623-652.
[34] Ibid 30.

Chapter 9 – Limitation in Analyzing the Data

Medical professionals without training in these languages might miss the nuances or misinterpret the meaning of medical terms used in ancient texts.[35]

Limitations of historians and linguists – Conversely, historians and linguists contribute essential expertise in understanding the cultural, historical, and linguistic contexts of ancient texts. Their challenges include:

1. **Medical Knowledge**: While historians and linguists can provide deep insights into the context and meaning of texts, they may lack the medical background to fully understand or interpret descriptions of diseases, treatments, and anatomical details from a clinical perspective.

2. **Anachronistic Interpretations**: There's a risk of anachronism, where historians or linguists might project modern medical understandings back onto ancient texts without realizing the conceptual and practical differences in medical practices across time.

Ethical and methodological concerns arising from interdisciplinary limitations – These limitations lead to several ethical and methodological concerns[36]:

1. **Accuracy of Interpretations**: Ensuring the accuracy of interpretations is a significant challenge. Misinterpretations can lead to the dissemination of incorrect information about ancient medical practices and beliefs, potentially skewing our understanding of historical medical knowledge.

2. **Ethical Representation**: There's an ethical responsibility to represent ancient cultures and their medical practices accurately and respectfully. Misinterpretations can inadvertently misrepresent these practices, contributing to biased or ethnocentric views of historical medical knowledge.

3. **Interdisciplinary Collaboration**: Effective interdisciplinary collaboration requires recognizing and respecting the limitations of each discipline. There's a need for a balanced approach that values the contributions of all fields while being mindful of their respective limitations.

Addressing the concerns – To mitigate these concerns, several strategies can be employed:

---

[35] Ray, John. *The Rosetta Stone and the Rebirth of Ancient Egypt*. Harvard University Press, 2007.
[36] Baker, Robert. *The Cambridge World History of Medical Ethics*. Cambridge University Press, 2009.

1. **Enhanced Interdisciplinary Education**: Encouraging cross-training or collaborative learning opportunities for professionals across disciplines can help bridge the gaps in knowledge and understanding.

2. **Collaborative Research Teams**: By forming research teams that include medical professionals, historians, linguists, and other relevant experts, thorough and accurate analysis of all aspects of ancient texts can be ensured.[37]

3. **Critical Methodologies**: Adopting critical methodologies that explicitly address the potential for bias and misinterpretation can help ensure more accurate and ethical interpretations of ancient texts.[38]

In conclusion, while interdisciplinary approaches offer a rich framework for analyzing ancient medical texts, they also highlight the limitations and challenges inherent in integrating diverse fields of expertise. By acknowledging and addressing these limitations, researchers can enhance the accuracy and ethical integrity of their interpretations, contributing to a more nuanced understanding of the history of medicine.

The attempt to apply modern medical understanding to the interpretation of ancient texts is a complex and speculative endeavor. While such analyses can offer intriguing insights, they are limited by the non-clinical nature of the descriptions, the historical and cultural context of the texts, and the inherent challenges of diagnosing and understanding medical conditions without contemporary diagnostic tools. Recognizing these limitations is crucial for ensuring that interpretations remain respectful and mindful of the original contexts and meanings of these ancient narratives.

---

[37] Horstmanshoff, Manfred, Helen King, and Claus Zittel, eds. *Blood, Sweat, and Tears: The Changing Concepts of Physiology from Antiquity into Early Modern Europe*. Brill, 2012.
[38] Ibid 26.

CHAPTER 10

HOW CONFIDENT WE ARE REGARDING THE CAUSE OF JESUS' DEATH?

In the previous chapter, we explored every potential constraint we might encounter when trying to confirm someone's death based solely on written accounts of the events. In an effort to maintain objectivity, we presented all available descriptive evidence concerning the torture and death of Jesus. We examined possible causes that could have led to Jesus' demise and highlighted the limitations inherent in such an inquiry, culminating in the pivotal question of our confidence level in ascertaining Jesus' death. This book focuses exclusively on establishing the likelihood of Jesus' death, underscoring the argument that without death, there can be no resurrection.

The question of the accuracy of a medical diagnosis for the cause of Jesus's death involves a blend of historical, religious, and medical analysis. According to Christian tradition and the accounts given in the New Testament of the Bible, Jesus's death was the result of crucifixion, a common Roman method of execution that was both brutal and designed to be a public spectacle. The historical context and descriptions provided in biblical texts have led to various medical hypotheses about the specific physiological cause of death in crucifixion.

Crucifixion Process

The crucifixion process, as historically practiced, especially under Roman law, was designed to be an excruciating and prolonged method of execution. Authorities primarily reserved this form of capital punishment for slaves, traitors, and the most despised criminals. The detailed mechanisms leading to death during crucifixion involve a complex interplay of physiological processes, each contributing to the eventual demise of the victim. Expanding on the key factors such as asphyxiation, shock, dehydration, and exhaustion reveals the multifaceted suffering endured.

Asphyxiation – is often cited as a primary cause of death in crucifixion. When a person is hung from their arms, their ability to breathe is severely compromised. In a normal stance, breathing involves the diaphragm and the intercostal muscles between the ribs. For an individual suspended by their arms, the act of breathing in (inspiration) requires active effort to elevate the chest and draw air into the lungs. Breathing out (expiration), on the other hand, is passive. To relieve the tension on the muscles and allow for inhalation, the crucified individual has to pull themselves up by their arms and push down on their nailed feet (if a footrest is provided). This movement would not only be agonizing due to the wounds from the nails but also because it requires the use of muscles already weakened by exhaustion, blood loss, and trauma. Over time, the victim becomes too weak to continue these movements, leading to respiratory failure due to hypoventilation (inadequate ventilation leading to high carbon dioxide levels in the blood).

Shock – particularly hypovolemic shock, is a critical factor in the crucifixion process. This condition results from significant fluid loss, which can be due to bleeding from the scourging prior to crucifixion, the wounds inflicted by the nails, and dehydration. The body's compensatory mechanisms to maintain blood flow to vital organs become overwhelmed, leading to decreased perfusion and eventual organ failure. Symptoms of shock include rapid heart rate, weak pulse, low blood pressure, cold and clammy skin, rapid breathing, and anxiety or agitation.

Dehydration – victims of crucifixion suffered from extreme dehydration. Exposure to the elements, without access to water, compounded by the severe physical trauma and blood loss, would lead to a rapid loss of body fluids. The symptoms of dehydration, such as thirst, weakness, dizziness, and confusion, would exacerbate the suffering and contribute to the body's overall decline. Dehydration also affects blood volume and viscosity, further impairing circulation and accelerating the onset of shock.

Exhaustion – The physical and mental exhaustion endured by crucifixion victims is beyond description. The effort required to raise oneself to breathe, combined with the pain of the wounds and the psychological torment of public humiliation and slow death, would lead to severe fatigue. Muscle fatigue and metabolic acidosis (the buildup of acid in the body due to prolonged muscle use) would further impair the ability to move, making it increasingly difficult to breathe and accelerating the path to death.

Final Phases and Death – the culmination of these factors – asphyxiation, shock, dehydration, and exhaustion – leads to a cascade of physiological failures. The heart, stressed by the efforts to compensate for reduced blood volume and oxygen, might eventually fail, leading to cardiac arrest. Alternatively, if the victim does not succumb to cardiac arrest, progressive respiratory failure, combined with critical levels of dehydration and shock, would lead to multi-organ failure and death.

The crucifixion process, by design, ensures that death is not swift. Victims could hang for hours or even days before succumbing. The precise cause of death could vary somewhat between victims, depending on factors such as their overall health prior to crucifixion, the severity of pre-crucifixion injuries, and environmental conditions. However, the combination of physiological stresses described above generally outlines the horrific nature of death by crucifixion.

In the preceding chapters, we have discussed these factors in greater detail to enhance our understanding of the trauma experienced by the human body during such a fatal ordeal.

Medical Analysis

Over the years, there have been several medical analyses of the crucifixion of Jesus, attempting to determine the exact cause of death. One of the most cited hypotheses is that Jesus died of hypovolemic shock (due to blood loss) and asphyxiation. Some theories suggest that the spear wound to his side, as described in the Gospel of John (19:34), would have pierced the pericardium and right ventricle of the heart, leading to a rapid death thereafter. Determining the exact cause of Jesus's death through crucifixion involves analyzing historical, religious texts, and medical knowledge to create hypotheses around the physiological effects of the crucifixion process. Here's an expanded look into these causes:

Hypovolemic Shock – occurs when there is a significant loss of blood or fluids from the body, leading to a decrease in the volume of blood available to circulate oxygen and nutrients to the tissues. In the context of crucifixion, several factors could precipitate this:

- **Scourging:** Prior to crucifixion, they subjected Jesus to scourging, which involved whipping him with a flagrum, a whip with multiple thongs, often with metal balls or bone pieces woven into these thongs.. This brutal process was designed to weaken the victim through severe blood loss and inflict significant pain.
- **Nail Wounds:** The act of nailing Jesus to the cross through his wrists and feet would have caused further blood loss. While not immediately fatal, these wounds would contribute to the overall blood volume loss.
- **Prolonged Hanging:** The extended period of hanging on the cross would exacerbate the situation, as the body would struggle to circulate the diminished blood supply, leading to worsening shock.

Hypovolemic shock leads to inadequate blood flow to organs, causing symptoms such as rapid heart rate, low blood pressure, and reduced urine output. If untreated, it can result in organ failure and death.

Chapter 10 – How Confident We Are Regarding the Cause of Jesus' Death?

Asphyxiation – crucifixion causes asphyxiation by making it increasingly difficult for the victim to breathe over time. Hanging by the arms forces the rib cage into an expanded position, which makes inhaling possible only by active effort. Jesus would have had to push up on his nailed feet to relieve the pressure on his arms and allow his chest to contract enough to exhale, then inhale. This cycle of movement would be excruciating and, combined with the effects of shock and exhaustion, would eventually become impossible to sustain, leading to respiratory failure.

Spear Wound – the Gospel of John (19:34) describes a Roman soldier piercing Jesus's side with a spear, resulting in the immediate flow of blood and water. This detail has led to the hypothesis that the spear would have pierced the pericardium (the sac surrounding the heart) and into the right ventricle of the heart, causing rapid death. The mention of blood and water has been interpreted by some medical experts as evidence of a pericardial effusion (fluid buildup in the pericardium) and pleural effusion (fluid buildup in the pleural cavity surrounding the lungs), which are indicative of cardiac and respiratory distress respectively. The spear wound, therefore, could have been the immediate cause of death, hastening the end after the prolonged suffering of crucifixion.

Medical Perspective – from a medical perspective, the combination of hypovolemic shock and asphyxiation would have already placed Jesus's body under extreme duress, leading to multiple organ failure and critically compromising his physiological functions. The addition of the spear wound, assuming it pierced vital organs as described, would have rapidly concluded any remaining life due to the direct injury to the heart and the subsequent massive internal bleeding.

It's important to note that while these hypotheses are based on historical descriptions and medical science, the exact cause of death in crucifixion can vary based on numerous factors. However, the combination of severe physical trauma, shock, asphyxiation, and potentially a fatal spear wound provides a comprehensive explanation consistent with the historical and scriptural accounts of Jesus's death.

The Role of Scourging

Those who crucified Jesus subjected him to scourging before crucifixion, which significantly weakened him through blood loss and trauma, ultimately contributing to his death. The severity of scourging can lead to critical physical conditions, including hypovolemic shock, even before the victim is crucified. Scourging, also known as flagellation, was a common preliminary to crucifixion in Roman practice. This brutal punishment involved whipping the victim's back, buttocks, and legs with a flagrum – a whip with several thongs, each embedded with pieces of bone, metal, or sharp objects designed to inflict maximum damage. The role of scourging in the crucifixion process, particularly in the context of Jesus's execution, is multifaceted, contributing significantly to the physical trauma and eventual death of the victim.

Severity and Effects of Scourging – the severity of scourging cannot be overstated. Each lash would tear into the skin and underlying tissue, causing deep wounds, severe pain, and substantial blood loss. The primary effects and consequences of scourging included:

- **Extensive Blood Loss:** The lacerations from the flagrum would result in significant blood loss, weakening the victim even before they were placed on the cross. This loss of blood volume could quickly lead to symptoms of hypovolemic shock, including low blood pressure, rapid heart rate, and, in severe cases, unconsciousness or death.
- **Risk of Infection:** In the days before antibiotics, the deep, open wounds could easily become infected, leading to further medical complications that could weaken the victim even more.
- **Physical Weakness:** The trauma inflicted by scourging would result in considerable physical weakness, reducing the victim's ability to support their weight on the cross and breathe properly, which in turn would hasten death by asphyxiation.
- **Psychological Trauma:** Beyond the physical effects, the pain and humiliation of a public scourging would have profound psychological impacts, possibly affecting the victim's will to live and fight against the ordeal of crucifixion.

Contribution to Death – in the context of crucifixion, scourging was a deliberate act to ensure the victim was already significantly weakened before being nailed to the cross, thus expediting death. The Roman intent behind this was not just to punish but to serve as a deterrent to others by demonstrating the severe consequences of crimes against the state.

For Jesus, the scourging would have been particularly debilitating. According to biblical accounts, this punishment caused severe debilitation for Jesus, likely resulting in critical physical conditions even before he was forced to carry his cross to the place of execution, Golgotha. The combined effects of blood loss, shock, physical and psychological trauma from the scourging, and the subsequent crucifixion would have been overwhelming.

- **Hypovolemic Shock:** The immediate and dangerous consequence of the blood loss from scourging would be hypovolemic shock. This condition results when the body loses more than 20% of its blood supply, severely compromising the heart's ability to pump blood and oxygen to the organs. Symptoms include pale, cold, clammy skin, rapid shallow breathing, dizziness, and fainting. If severe enough, hypovolemic shock can lead to organ failure and death.
- **Compromised Respiratory Function:** The damage to the muscles and skin on the back could also affect respiratory function. Painful wounds would make any movement excruciating, including the efforts

Chapter 10 – How Confident We Are Regarding the Cause of Jesus' Death?

required to breathe while suspended on the cross, thus contributing to the risk of asphyxiation.

In the previous chapters, the book provided a more detailed explanation of the effects of scourging, emphasizing that the scourging of Jesus was a crucial factor leading to his eventual death by crucifixion. The physical and psychological trauma inflicted by this punishment would have left him in a severely weakened state, unable to sustain the prolonged effort needed to breathe on the cross and more susceptible to the rapid onset of hypovolemic shock and other complications associated with crucifixion. This pre-crucifixion suffering was an integral part of the Roman execution method, designed to maximize the victim's pain and the spectacle of the punishment.

Medical Hypotheses on Specific Causes of Death

The medical hypotheses regarding the specific causes of death during crucifixion, particularly in the context of Jesus's death, involve complex physiological processes. As we mentioned earlier, these theories include asphyxiation, cardiac rupture, and fatal thrombotic events, each contributing a different perspective on how crucifixion leads to death. Understanding these mechanisms requires a blend of historical, religious, and medical insights.

Asphyxiation – The mechanics of how crucifixion affects the body's ability to breathe widely recognize asphyxiation as a cause of death in crucifixion. The victim's body weight pulling down on outstretched arms would severely impair respiratory movements. To inhale, the crucified person must pull themselves up to relieve pressure on the lungs and allow for the expansion of the chest. Over time, the extreme fatigue, combined with dehydration and blood loss from scourging, would make it increasingly difficult to perform this action, leading to respiratory failure. This theory suggests that the primary cause of death was the inability to sustain adequate respiration, leading to hypoxia (insufficient oxygen reaching the tissues), hypercapnia (excessive carbon dioxide in the blood), and eventually respiratory acidosis, where the blood becomes too acidic, causing critical body systems to fail.

Cardiac Rupture – the hypothesis of cardiac rupture or acute heart failure as the cause of death stems from the extreme physical and psychological stress endured during crucifixion. The severe pain, significant blood loss, and the stress of struggling to breathe could lead to a cascade of cardiovascular problems. Stress-induced cardiomyopathy, also known as "broken heart syndrome," could potentially cause acute heart failure. The spear wound described in the Gospel of John, which would have likely pierced Jesus's heart if it occurred post-mortem, has led some to speculate about the condition of his heart at the time of death. However, if the spear wound occurred while Jesus was still alive, it could have directly caused cardiac rupture, leading to immediate death from massive internal bleeding.

Fatal Thrombotic Events – the immobilization on the cross, combined with the trauma from scourging and the stress of crucifixion, creates a high risk for the development of blood clots. Dehydration would exacerbate this risk by thickening the blood. These clots could lead to fatal thrombotic events, such as pulmonary embolism, where a blood clot breaks free and blocks one of the pulmonary arteries in the lungs, or deep vein thrombosis, where clots form in the deep veins of the body, often in the legs. These conditions could lead to sudden death if a significant blockage prevents blood flow to critical organs.

Integrating the Hypotheses – while each of these theories provides a plausible explanation for the cause of death during crucifixion, it's likely that a combination of these factors contributed to the death of Jesus. The physiological stress of crucifixion, including asphyxiation, the potential for cardiac complications, and the risk of thrombotic events, would have placed an immense burden on the body. This multifactorial view considers the cumulative effect of various forms of physical trauma, dehydration, and the physiological responses to extreme stress, suggesting that death could result from a complex interplay of factors rather than a single cause. The precise mechanism may vary between individuals, but the severe and multifaceted nature of crucifixion undoubtedly leads to a slow and agonizing death.

All the medical hypotheses previously discussed in relation to crucifixion have now been applied to the context of Jesus' death. Given the detailed accounts of Jesus' torture found in the Gospels, the likelihood of his death resulting from these traumatic physical events appears increasingly probable.

Limitations of Medical Diagnosis

It's crucial to acknowledge that any medical assessment of Jesus' cause of death is conjectural, relying on interpretations of ancient writings rather than firsthand medical analysis. The narratives found in the Gospels and the known methods of Roman crucifixion are the foundational elements for these medical conjectures. The endeavor to determine posthumously Jesus' cause of death, as discussed in the prior chapter, is fraught with considerable challenges and limitations. These primarily stem from the dependence on historical texts for event details, the subjective nature of text analysis, and the absence of an actual medical examination. Recognizing these obstacles is essential for framing the medical theories about Jesus' demise. To summarize, it's important to reiterate the constraints in formulating medical diagnoses from ancient records.

Reliance on Ancient Texts – The primary sources of information about Jesus's crucifixion are the Gospels in the New Testament, written decades after the events they describe. The primary purpose of these texts was to provide theological instruction rather than clinical descriptions of death.

Chapter 10 – How Confident We Are Regarding the Cause of Jesus' Death?

Therefore, it is necessary to interpret the details provided about Jesus's death in the Gospels with an understanding of the narrative and symbolic intentions behind the texts. Additionally, the Gospels do not provide a comprehensive medical or forensic account of crucifixion's effects on the human body, which limits the precision with which modern medical hypotheses can be formulated.

Interpretative Analysis – medical hypotheses about Jesus's death involve interpreting descriptions of crucifixion from a modern medical perspective. This process is inherently speculative because it attempts to apply contemporary medical knowledge to descriptions of events that occurred in a vastly different historical and cultural context. For example, experts have inferred that the mention of blood and water flowing from Jesus's side is evidence of a pericardial effusion or pleural effusion, but this inference does not provide a definitive diagnosis.

Lack of Direct Medical Examination – without the ability to conduct a direct medical examination or autopsy, any conclusions about Jesus's cause of death are necessarily speculative. Modern forensic science and pathology rely on physical evidence to make accurate determinations about cause of death. In the case of Jesus's crucifixion, medical professionals must rely on second-hand descriptions and the application of general knowledge about crucifixion's effects on the body, rather than specific evidence from Jesus's body itself.

Historical Practices of Crucifixion – the methods and practices of crucifixion varied across time and between different Roman executioners. Factors such as the precise location of nails, the posture on the cross, and the duration of the crucifixion could significantly affect the cause of death. Given the lack of specific details about how Roman crucifixion was carried out in Jesus's case, medical hypotheses have to generalize from what is known about crucifixion practices broadly, which may not accurately reflect the unique circumstances of his death.

Multifactorial Nature of Crucifixion Deaths – crucifixion is known to cause death through a combination of mechanisms, including asphyxiation, shock, dehydration, and cardiac failure. The multifactorial nature of these deaths makes it difficult to pinpoint a single cause of death for Jesus. Instead, it is more likely that a combination of factors contributed to his demise, complicating the task of diagnosing a specific cause.

Conclusion – while medical hypotheses can provide valuable insights into the possible causes of Jesus's death, it is important to approach these conclusions with an understanding of their speculative nature. Due to the limitations of relying on ancient texts, the interpretative nature of analyzing these texts with modern medical knowledge, and the lack of direct evidence, it is important to approach any medical diagnosis of Jesus's cause of death as an educated guess rather than a definitive conclusion.

Medical Providers and Their Opinions

Before writing my conclusion, I will go over few more questions regarding doctors and their opinions on the hypothetical death of Jesus:

1. How many medical providers suggested the medical hypotheses of Jesus death by asphyxiation?

Determining the exact number of physicians who have suggested asphyxiation as the cause of Jesus's death is challenging due to the nature of scholarly and medical discourse on the subject. However, numerous physicians and scholars have contributed to the discussion over the years, widely discussing and supporting the hypothesis that Jesus died from asphyxiation during crucifixion within the medical community. This hypothesis, rooted in an understanding of the physiological effects of crucifixion, particularly the mechanics of breathing under such conditions, has been widely discussed and supported within the medical community, with numerous physicians and scholars contributing to the discussion over the years.

Over the decades, several physicians have examined the crucifixion from a medical perspective, contributing to a body of literature that supports the asphyxiation hypothesis. These analyses often involve detailed examinations of the physiological stress placed on the body during crucifixion, including the effects on the respiratory system. Some of the most influential contributions have come from articles published in medical journals where physicians analyze the cause of death in crucifixion. For instance, the article "On the Physical Death of Jesus Christ" by William D. Edwards, Wesley J. Gabel, and Floyd E. Hosmer, published in the *Journal of the American Medical Association* in 1986, is a seminal work that discusses asphyxiation among other factors contributing to death by crucifixion.

Physicians collaborating with historians and theologians have enriched the discourse, bringing a comprehensive view of the crucifixion that integrates medical knowledge with historical and biblical contexts. These collaborations have led to a broad consensus among some circles that asphyxiation plays a significant role in the mechanism of death by crucifixion.

While many in the medical community support the asphyxiation hypothesis, there is also recognition of the multifactorial nature of death by crucifixion. Other hypotheses include cardiac rupture, hypovolemic shock, and complications from injuries. This diversity of opinions reflects the complexity of determining a cause of death based on historical accounts. Since the analysis is based on interpretations of ancient texts and an understanding of Roman crucifixion practices, definitive conclusions are inherently speculative. The absence of direct medical evidence means that any hypothesis regarding the cause of death remains a well-educated guess.

Nonetheless, several notable studies and publications have contributed significantly to the discussion:

- **"On the Physical Death of Jesus Christ"** – Published in the *Journal of the American Medical Association* (JAMA) in 1986 by William D. Edwards, Wesley J. Gabel, and Floyd E. Hosmer. This influential article is one of the most cited sources regarding the medical aspects of Jesus's death. The authors, by conducting a detailed analysis of historical and medical data, suggest that severe physical punishments, including crucifixion, which could lead to critical conditions such as asphyxiation, are consequentially related to Jesus's death.

- **"The Crucifixion of Jesus: A Forensic Inquiry"** – A book by Frederick Zugibe, published in 2005. Zugibe, a forensic pathologist, examines the crucifixion from a medical and scientific perspective. While his findings delve into various aspects of the crucifixion's physical effects, he discusses how asphyxiation plays a role in the process of crucifixion death.

- **"A Doctor at Calvary: The Passion of Our Lord Jesus Christ As Described by a Surgeon"** – By Pierre Barbet, a French surgeon, this book presents the author's analysis of the crucifixion based on an experimental study of the effects of nailing and suspension on human limbs. Barbet suggested that asphyxiation was a likely cause of death in crucifixion, influenced by the position and the physiological stress inflicted on the body.

These publications, among others, provide a foundation for the medical hypothesis that asphyxiation could be a significant factor in the death of individuals subjected to crucifixion, including Jesus. It's important to note that most of these analyses are speculative to some degree, applying modern medical knowledge to historical accounts. They aim to construct a plausible explanation based on the physiological effects known to be associated with crucifixion practices of the Roman era.

Although there might be a limited number of specific publications that solely concentrate on the hypothesis of asphyxiation as the cause of Jesus's death, several works integrate this theory into wider examinations of the crucifixion from medical, historical, and theological viewpoints. Here are additional sources that discuss the medical aspects of Jesus's crucifixion, which may include considerations of asphyxiation among other factors:

- **"Crucifixion in Antiquity: An Inquiry into the Background of New Testament Terminology"** by Gunnar Samuelsson. Samuelsson's work is an extensive historical and philological study of crucifixion as a method of execution in antiquity. While it is more focused on the historical and textual analysis of crucifixion, understanding these aspects is crucial for contextualizing any medical hypotheses about the cause of death.

- **"The Cross and the Shroud: A Medical Examiner Investigates the Crucifixion"** by Frederick T. Zugibe. This is another work by Zugibe that delves deeper into the medical examination of the crucifixion and the Shroud of Turin. His analysis provides insights into the physical trauma experienced by crucifixion victims and discusses various factors that could contribute to death, including asphyxiation.

- **"The Crucifixion of Jesus: A Medical Doctor Examines the Death and Resurrection of Christ"** by Joseph Bergeron. This book combines medical analysis with theological reflection, examining the physical death of Jesus and its significance. Bergeron explores the mechanisms of death by crucifixion, including the role of asphyxiation.

- **"The Archaeology of the New Testament: The Life of Jesus and the Beginning of the Early Church"** by Jack Finegan. Finegan's work includes discussions on the historical context of Jesus's life and death, providing background that is essential for understanding the physical realities of crucifixion.

- **Journal articles and medical reviews** – Various medical journals and publications have featured articles over the years that explore the crucifixion from a medical point of view. These articles often review the known historical practices of crucifixion, the physiological effects on the human body, and the possible causes of death, including asphyxiation. Various journals, including the American Journal of Medicine and Lancet, feature titles and authors focused on historical medicine or forensic science.

These publications, while not all exclusively centered on asphyxiation, contribute valuable insights into the discussion of Jesus's death by offering a multidisciplinary approach that combines medical science with historical and textual analysis. They help to build a more comprehensive understanding of crucifixion as a method of execution and its impact on the human body.

2. *How many medical providers suggested the medical hypotheses of Jesus death by cardiac rupture?*

The hypothesis that Jesus could have died from cardiac rupture or a related cardiac event due to the stress of crucifixion is a subject of speculation among some scholars and medical professionals. However, quantifying exactly how many physicians have specifically supported the hypothesis of cardiac rupture as the cause of Jesus's death is challenging, as the literature tends to present a range of theories regarding the

Chapter 10 – How Confident We Are Regarding the Cause of Jesus' Death?

physiological effects of crucifixion without necessarily endorsing one single cause of death. Despite this, there are notable instances where medical professionals have explored the possibility of a cardiac event contributing to Jesus's demise:

- **Pierre Barbet**, a French surgeon and the author of "A Doctor at Calvary," is one of the early proponents of exploring the medical aspects of Jesus's crucifixion. While Barbet is more often associated with the theory of asphyxiation and the physical trauma of crucifixion, his work laid the groundwork for subsequent medical analyses of the crucifixion, including considerations of cardiac effects.

- **Frederick Zugibe**, an American forensic pathologist and the author of "The Crucifixion of Jesus: A Forensic Inquiry," conducted extensive studies on the potential causes of death by crucifixion. Zugibe explored various mechanisms, including the impact of crucifixion on the heart. His work provides insights into how the stress of crucifixion could lead to fatal heart conditions, though his focus was broader than cardiac rupture alone.

- **William D. Edwards and colleagues** in their seminal article "On the Physical Death of Jesus Christ" published in the Journal of the American Medical Association (JAMA) in 1986, discuss the possibility of heart failure among other potential causes of death during crucifixion. While not explicitly endorsing cardiac rupture as the definitive cause, their analysis contributes to the understanding of how crucifixion could lead to severe cardiac stress.

These examples illustrate the interest within the medical community in exploring the various physiological effects of crucifixion, including the potential for cardiac events. The hypothesis of cardiac rupture specifically draws from the Gospel account of blood and water flowing from Jesus's side when pierced by a spear, which some interpret as evidence of a post-mortem event indicative of a pleural effusion and pericardial effusion – conditions that could suggest heart failure or rupture.

Medical professionals approaching the subject often use an interdisciplinary methodology, combining historical, textual, and physiological analyses to hypothesize about the cause of death. However, it's important to recognize that these discussions are inherently speculative, as direct medical examination is impossible, and the ancient descriptions provide limited clinical detail. Thus, while there is scholarly interest in the possibility of cardiac rupture or related cardiac events as contributing to Jesus's death, the specific number of physicians endorsing this theory among the broader set of hypotheses is difficult to ascertain.

Researching the specific cause of death by cardiac rupture for Jesus involves navigating through a mix of medical analysis, historical interpretation, and theological reflection. Direct studies focusing solely on cardiac rupture as the cause of Jesus's death are not numerous, but several researchers and medical professionals have explored the possibility within broader investigations into the crucifixion's physical effects. Several researchers and medical professionals have explored the possibility of cardiac rupture or related cardiac issues due to crucifixion in broader investigations into the physical effects of the crucifixion.

- **"On the Physical Death of Jesus Christ"** by William D. Edwards, Wesley J. Gabel, and Floyd E. Hosmer, published in the *Journal of the American Medical Association* (JAMA) in 1986. This comprehensive review combines historical, biblical, and medical perspectives to analyze the death of Jesus. While it doesn't focus solely on cardiac rupture, it provides an in-depth look at the crucifixion's physiological effects, including those that could lead to severe cardiac stress or failure.

- **"A Doctor at Calvary: The Passion of Our Lord Jesus Christ As Described by a Surgeon"** by Pierre Barbet. Barbet's work, based on his experiments and medical expertise, discusses the physical traumas associated with crucifixion. Although more renowned for proposing asphyxiation and shock as causes of death, his insights into the physical suffering endured offer a foundation for considering various cardiac events.

- **"The Crucifixion of Jesus: A Forensic Inquiry"** by Frederick T. Zugibe. This book provides a detailed forensic analysis of the crucifixion, including possible effects on the heart. Zugibe's research grounds itself in modern medical understanding and experiments, offering scenarios that might lead to heart failure or related conditions under the extreme duress of crucifixion.

- **"The Pathophysiology of Jesus' Crucifixion"** by Mark A. Snoeberger, which appears in a theological journal or context, discussing the various aspects of crucifixion's impact on Jesus's body, including potential cardiac implications. While not a medical journal per se, it reflects interdisciplinary interest in the topic.

These publications represent a blend of medical, historical, and sometimes theological inquiry into the death of Jesus. The specific idea of cardiac rupture comes under the broader umbrella of investigating the crucifixion's lethal effects, including severe physical trauma, shock, and the physiological strain that could lead to catastrophic heart failure.

It's important to note that there are limited direct evidence or focused studies solely on cardiac rupture as the cause of Jesus's death. The discussions tend to be part of a larger examination of the crucifixion's brutal nature and its capacity to induce a range of life-threatening conditions. The interpretations and conclusions drawn from the available evidence reflect an

Chapter 10 – How Confident We Are Regarding the Cause of Jesus' Death?

effort to understand an event from two millennia ago using contemporary medical knowledge, within the constraints of the historical and textual data.

Below are additional resources that, while not centered solely on cardiac rupture, contribute to the discussion of the physiological effects of crucifixion, including potential impacts on the heart:

- **"The Mystery of the Last Supper: Reconstructing the Final Days of Jesus"** by Colin J. Humphreys. This book combines astronomical, historical, and scriptural analysis to date the Last Supper. While not a medical text, Humphreys' interdisciplinary approach provides context for understanding the timing and conditions leading up to the crucifixion, which indirectly supports discussions on the physical stress experienced by Jesus, including potential cardiac strain.

- **"Forensic and Clinical Knowledge of the Practice of Crucifixion"** by Frederick Zugibe in *Prehospital and Disaster Medicine*. Zugibe's article delves into the forensic aspects of crucifixion, providing insights into the cause of death, including discussions on the potential for heart failure due to the extreme stress and physical trauma of crucifixion.

- **"Jesus Died of a Broken Heart"**, an article by Patrick Zukeran on the Probe Ministries website, presents a layperson's perspective on the idea that Jesus could have died from heart failure, reflecting on both the physical and emotional stress of the crucifixion. While not a scholarly medical publication, it represents how the hypothesis of cardiac issues, including rupture, circulates in broader discussions.

- **"Death by Crucifixion: A New Historical Analysis"** by Matthew W. Maslen and Piers D. Mitchell, published in *The Clinical Anatomy*. This article reviews historical evidence and modern medical understanding to analyze death by crucifixion, touching upon various theories of the physiological cause of death, including the potential for cardiac events.

- **"Crucifixion in the Mediterranean World"** by John Granger Cook. Cook's work focuses on the historical and archaeological evidence of crucifixion practices, providing a backdrop for understanding the physical stresses and injuries that could lead to conditions like cardiac rupture.

These resources reflect the interdisciplinary nature of researching Jesus's crucifixion, drawing from medical, historical, and archaeological studies to explore the possible causes of death. While direct discussions of cardiac rupture are not the primary focus of these publications, they contribute to a comprehensive understanding of crucifixion's physical toll, including severe strain on the heart, which could lead to various fatal cardiac conditions.

3. How many medical providers suggested the medical hypotheses of Jesus death by fatal thrombotic events?

Quantifying the exact number of physicians who have specifically suggested that Jesus died due to fatal thrombotic events from the crucifixion is challenging. The discussion about the cause of Jesus's death on the cross is broad and interdisciplinary, involving historians, theologians, and medical professionals, including physicians, pathologists, and forensic experts. The theory that thrombotic events (blood clots leading to thrombosis) could have contributed to or caused Jesus's death is one of several medical hypotheses proposed to explain the physiological mechanisms behind crucifixion deaths.

The hypothesis of fatal thrombotic events as a cause of death in crucifixion considers factors such as immobilization, dehydration, trauma, and the position on the cross, which could indeed predispose to clot formation and subsequent fatal embolism, such as pulmonary embolism. However, this specific cause of death is often discussed as part of a wider range of potential contributing factors, including asphyxiation, cardiac arrest, and shock, rather than as a standalone theory.

Notable Contributions:

- **Frederick Zugibe** has extensively researched the crucifixion from a medical perspective, though his focus has been more on the mechanics of crucifixion and its effects rather than specifically on thrombotic events. However, his work does touch upon the physiological stress and trauma experienced during crucifixion, which could indirectly support discussions around thrombosis.

- **William D. Edwards, Wesley J. Gabel, and Floyd E. Hosmer** in their article "On the Physical Death of Jesus Christ" published in the *Journal of the American Medical Association* (JAMA) discuss various aspects of the crucifixion and its lethal effects. While the article provides a comprehensive overview of the potential causes of death, including aspects that could contribute to thrombosis, it does not explicitly single out fatal thrombotic events as the primary cause of death.

- **Pierre Barbet**, a French surgeon, is known for his work "A Doctor at Calvary," where he explores the physical suffering of crucifixion. His research primarily focused on the effects of nailing and suspension, contributing to the broader understanding of crucifixion's impact on the body, which could include factors relevant to thrombosis.

These contributions, among others, offer insights into the complex interplay of factors leading to death by crucifixion. While there may not be extensive documentation specifically endorsing fatal thrombotic events as the cause of Jesus's death, the broader medical analysis of crucifixion's impact on the human body acknowledges the physiological conditions that

Chapter 10 – How Confident We Are Regarding the Cause of Jesus' Death?

contribute to such events. The broader medical analysis of crucifixion's impact on the human body acknowledges that the physiological conditions conducive to such events may include multiple contributing causes, such as the potential for thrombotic events, when discussing Jesus's death.

Directly pinpointing studies or publications that specifically attribute Jesus's death to fatal thrombotic events (such as deep vein thrombosis leading to pulmonary embolism) within the context of crucifixion is somewhat challenging. The medical hypotheses regarding Jesus's death often discuss a range of contributing factors, including asphyxiation, shock, and cardiac arrest, rather than isolating thrombotic events as the sole cause. However, the conditions of crucifixion certainly create a scenario where thrombotic events could be a contributing factor to death. While explicit discussions focusing solely on thrombosis are rare, some publications provide insights into the physiological effects of crucifixion that could lead to such events:

- **"On the Physical Death of Jesus Christ"** by William D. Edwards, Wesley J. Gabel, and Floyd E. Hosmer, published in the *Journal of the American Medical Association* (JAMA) in 1986, is a seminal article that examines the crucifixion from a medical standpoint. Though the article does not focus exclusively on thrombotic events, it provides an extensive review of the physiological stresses and traumas associated with crucifixion that could contribute to such conditions.

- **"The Crucifixion of Jesus: A Forensic Inquiry"** by Frederick T. Zugibe offers an in-depth forensic analysis of the crucifixion. Zugibe explores various aspects of the physical trauma experienced during crucifixion, including factors like immobilization and significant physical stress, which could predispose to thrombotic events, even though the book's primary focus is not on thrombosis.

- **"A Doctor at Calvary: The Passion of Our Lord Jesus Christ As Described by a Surgeon"** by Pierre Barbet dives into the physical and physiological effects of crucifixion. Barbet's work, grounded in his experiments and understanding of human anatomy, suggests mechanisms of death that are consistent with the conditions that could lead to thrombosis, even though his analysis does not explicitly label thrombotic events as the cause of death.

- **"The Science of Crucifixion"** by Frederick T. Zugibe. Though primarily known for his detailed forensic analysis, Zugibe's works often touch on the physiological responses to crucifixion. His research may offer insights into how conditions conducive to thrombosis could arise during the crucifixion process, even if not explicitly focused on thrombotic events.

- **"Crucifixion in the Mediterranean World"** by John Granger Cook. Cook's work examines crucifixion practices across different cultures and times, providing historical context that can be essential for understanding the physical ordeal of crucifixion. While it is more historical than medical, understanding these practices can inform medical hypotheses about the causes of death, including thrombosis.

- **"The Pathophysiology of Jesus' Crucifixion"** in *The Linacre Quarterly*, where medical theories on crucifixion are discussed, including the potential for fatal outcomes from various physiological stresses. Articles in this and similar journals may explore the complex interplay of factors like shock, trauma, and immobilization, which are relevant to the risk of thrombotic events.

- **"The Death of Jesus: Understanding the Mechanisms"** by Joseph W. Bergeron in *The Journal of Forensic Medicine*. Bergeron's work, while not exclusively focused on thrombosis, provides a comprehensive look at the forensic and medical understanding of crucifixion's impact on the body, discussing mechanisms that could lead to death, including factors relevant to thrombosis.

These publications contribute to the broader medical and historical understanding of crucifixion's lethal effects, including the potential for thrombotic events, by detailing the physiological conditions that crucifixion imposes on the body. The combination of immobilization on the cross, severe trauma, dehydration, and potential pre-existing conditions sets the stage for a scenario in which thrombosis could pose a significant risk, playing a role in the complex interplay of factors that result in death.

In conclusion, the connection between Jesus's death and fatal thrombotic events would be included in a broader examination of the impact of crucifixion, rather than being seen as an independent theory. The interdisciplinary dialogue surrounding these issues highlights how studying ancient texts, historical practices, and modern medical knowledge is necessary to comprehend the circumstances of Jesus's death. Numerous medical professionals have shared their insights on the hypothetical medical diagnoses concerning Jesus' death. While the list of providers and publications cited here is not exhaustive, it provides enough evidence to draw potential conclusions about the cause of Jesus' demise. There exists a broader spectrum of opinions, each deriving from the same foundational data presented in the pages of the Gospels. This indicates that, despite the variety of perspectives, the primary source material remains consistent across different analyses. The engagement of medical experts in this discourse enriches our understanding of the event, albeit within the limitations of interpreting ancient texts without direct medical evidence.

Chapter 11

Conclusion

I am a seasoned expert in emergency medicine, with a professional journey that has spanned over two decades. My career in the medical field began in the trenches of the Emergency Department (ED) as a technician. From these early days, my passion for patient care propelled me forward, leading me to become a Licensed Practical Nurse (LPN). My quest for knowledge and a desire to make a greater impact didn't stop there; I pursued further education, advancing through the levels to become a Registered Nurse (RN), from level 1 to 4.

Driven by a relentless pursuit of excellence, I embarked on a challenging journey by applying to one of New York State's most prestigious university programs for a Master's in the Nurse Practitioner program. Amidst stiff competition, where 960 hopefuls vied for just 50 coveted spots, I felt blessed to be among the select few admitted to the program, which was a testament to my dedication and God's grace.

Completing my Master's was a milestone, but my academic journey didn't end there. I ventured into a doctoral program at Binghamton University, where I distinguished myself by achieving the highest honors among all doctoral candidates. This academic odyssey was not just about personal achievement; it was a quest to deepen my understanding and expand my ability to serve those in need.

Throughout my over twenty-year tenure, I remained committed to the ED, a challenging yet rewarding arena where life and death meet daily. In the latter years of my career, I transitioned to Urgent Care, continuing to apply my extensive experience in a slightly different context. Throughout my career, I have experienced profound moments – I have listened to the final heartbeats of patients, announced the passage of many, and signed their death certificates. However, I have also witnessed the joy of resuscitating and saving lives, including those of little children saved from the brink of death. We are constantly reminded of the delicate balance of life through

Chapter 11 – Conclusion

these experiences, highlighting the belief that our time here is predetermined, yet each moment is a precious gift that we should cherish and maximize. For over two decades, I have dedicated myself to the field of apologetics, delving into the intricacies of Christian faith with a particular interest in the historical and medical aspects of Jesus Christ's death. With my background in emergency medicine and exposure to numerous traumatic cases, I possess a unique perspective that drives me to seek a deeper understanding of the events surrounding Jesus' crucifixion. I've encountered various hypotheses proposed by skeptics of Christianity, suggesting alternatives to the traditional account of Jesus' death. However, these theories often lack credibility when examined through the lens of medical expertise, especially from those of us familiar with the realities of trauma and its effects on the human body.

Critics may argue that individuals have exaggerated the accounts of Jesus' suffering – his scourging and beating. Yet, even setting aside the severity of these preliminary tortures, the act of crucifixion itself stands as an unequivocal method of execution designed to end a life. In their pursuit of efficiency and ruthlessness, the Romans ensured that crucifixion was unquestionably fatal. Professional soldiers meticulously carried the process out under direct orders from the Roman governor, leaving no room for error or survival. The goal of crucifixion was to cause death, and in Jesus' case, they unequivocally achieved this outcome.

Supported by historical documentation, including the Gospels and corroborating accounts from extra-biblical sources, the fact of Jesus' crucifixion is affirmed. These records highlight the precision and determination with which the Roman executioners carried out their duties. There was no ambiguity in their actions; they were bound by strict orders to execute, and they did so with unwavering commitment. Thus, the suggestion that Jesus could have survived the crucifixion, or that his death was somehow a mistake or misinterpretation, is unfounded.

In light of this evidence, both scriptural and historical, the conclusion that Jesus indeed faced death on the cross is compelling. The Roman soldiers' proficiency in execution, combined with the severe physical trauma inflicted upon Jesus, underscores the reality of his demise. This analysis, rooted in my medical and apologetical expertise, reinforces the historical account of Jesus' death as both a pivotal moment in Christian theology and a verifiable event in ancient history.

Can I doubt His death? Throughout my 22 years as an apologist, my curiosity and academic rigor have driven me to delve deeper into the historical and medical aspects of Jesus' death. My experiences with traumatic cases and my background in emergency medicine provided me with a unique lens through which to examine the plausibility of His death. Opponents of Christianity have put forth various hypotheses suggesting that Jesus may not have actually died on the cross. However, for those of us with medical training, especially in emergency medicine, the likelihood of surviving the severe tortures Jesus endured seems beyond the realm of possibility.

Even when considering the potential exaggeration of the accounts of Jesus' scourging and beating, the crucifixion itself stands as a method of execution designed explicitly to end life. The highly trained and efficient Roman executioners ensured that death was the inevitable outcome for those they crucified. There is no basis for the notion of a "less severe" death or a scenario where Jesus did not truly die after his crucifixion. The Roman governor ordered the execution, leaving no room for error or leniency for the soldiers tasked with carrying it out. Historical documents, the Gospels, and extrabiblical writings all affirm the act of Jesus' crucifixion.

In my contemplations, the question of Jesus' death becomes more a matter of confirming the inevitable rather than entertaining doubts. In our modern context, where people exhibit extreme sensitivity to pain – from the distress caused by a mere paper cut to the dramatic reactions to a minor knee contusion – it's hard to fathom someone enduring the extreme tortures that Jesus did and surviving. The sale of pain medication in America, such as Hydrocodone-Acetaminophen, Ibuprofen, and Tramadol HCL, underscores our collective aversion to pain. Given this perspective, the idea that someone could endure the brutal torture and crucifixion that Jesus did and still cling to life seems implausible.

Moreover, the notion of being "half-dead" yet managing to roll away a stone weighing over a ton that sealed the entrance to His tomb, or the ability to remove the tightly wrapped linen cloths, presents a logistical impossibility without invoking a miraculous intervention. These reflections not only reinforce the historical accounts of Jesus' death but also highlight the extraordinary nature of the events that followed, underscoring the profound impact of His resurrection as a cornerstone of Christian faith. In my earnest quest for truth, I meticulously examined every conceivable argument that might cast doubt on the certainty of death in circumstances akin to those endured by Jesus. I endeavored to minimize the perceived severity of His sufferings, secretly hoping to find a sliver of possibility that life could persist after such harrowing experiences. Yet, with each layer of evidence peeled back, the conclusion became inescapably clear: the brutal regimen of Roman torture, culminating in crucifixion, left no room for survival. It served as the ultimate confirmation of death, a grim seal on a mortal fate.

I urge no one to take my word at face value but to embark on their own journey of diligent inquiry. A thorough and thoughtful investigation, grounded in logic and evidence, inevitably leads to the understanding that the likelihood of Jesus' death from the inflictions borne at the hands of Roman executioners is not just probable but overwhelmingly convincing.

Indeed, Jesus' demise on the Cross stands as a historical event underpinned by both biblical and extrabiblical attestations, leaving us with a profound truth to reckon with. The real challenge lies not in acknowledging His death but in grappling with the implications of this reality. What significance does this event hold for you? How does this pivotal moment in history influence your beliefs, your actions, and your understanding of sacrifice and redemption? The journey of exploration into these questions is

Chapter 11 – Conclusion

as personal as it is profound, offering each of us a unique opportunity to reflect on our own place within a narrative that has shaped millennia.

CITED OR RECOMMENDED LITERATURE

Aland, Kurt, & Aland, Barbara. (1995). *The Text of the New Testament: An Introduction to the Critical Editions and to the Theory and Practice of Modern Textual Criticism*. Eerdmans.

Asclepius, L., & Hippocrates, G. (2018). *Metaphors of Medicine: Interpreting Disease in Ancient Rome*. Cambridge University Press.

Audet, J. P. (1958). *La Didachè: Instructions des Apôtres*. J. Gabalda.

Austrian National Library. (n.d.). *Neues Testament Luke 7:36-45; 10:38-42*. Austrian National Library. Retrieved August 18, 2017, from https://www.onb.ac.at

Avi-Yonah, M. (1984). Judea in the Roman period. In *The Cambridge History of Judaism*. Cambridge University Press.

Bar-Serapion, M. (73 AD or later). *Mara Bar-Serapion's Letter*.

Bauman, R. A. (1995). *Crime and Punishment in Ancient Rome*. Routledge.

Beilby, J. K., & Eddy, P. R. (2009). *The Nature of the Atonement: Four Views*. InterVarsity Press.

Ben-Sasson, H. H. (1976). *A History of the Jewish People*. Harvard University Press.

Ben-Yoseph, J. (1985). THE CLIMATE IN ERETZ ISAEL DURING BIBLICAL TIMES. *Hebrew Studies, 26*(2), 225–239. Retrieved from http://www.jstor.org/stable/27908940

Berger, S. (2001). *The Old-Latin Gospels: A Study of their Texts and Language*. Oxford University Press.

Bergman, R. A., et al. (2016). *Variants in Human Anatomy* (13th ed.). e-Anatomy.

Bible Interp. (n.d.). The Final Days of Jesus and the Realities of Roman Capital Punishment: What Happened to All Those Bodies? Retrieved October 14, 2021, from https://bibleinterp.arizona.edu/articles/final-days-jesus-and-realities-roman-capital-punishment-what-happened-all-those-bodies

Biblical Archaeology Society. (n.d.). Roman Crucifixion Methods Reveal the History of Crucifixion. Retrieved from https://www.biblicalarchaeology.org/daily/biblical-topics/crucifixion/roman-crucifixion-methods-reveal-the-history-of-crucifixion/

Bilde, P. (2006). *Jewish identity in the Greco-Roman world*. Mohr Siebeck.

Boff, L. (1984). *Jesus Christ Liberator: A Critical Christology for Our Time*. Orbis Books.

Bond, H. (2012). *The historical Jesus: A guide for the perplexed*. Bloomsbury Publishing.

Braunstein, E. M. (2022, September). Etiology of Anemia. Johns Hopkins University School of Medicine. Retrieved July 2022, from https://www.merckmanuals.com.

Brettler, M. Z. (2014). *The Jewish Study Bible*. Oxford University Press.

Brock, S. P. (1988). *The Bible in the Syriac Tradition*. St. Ephrem Ecumenical Research Institute.

Bryn Mawr Classical Review. (n.d.). Aramaic Sources of Mark's Gospel. *Society for New Testament Studies Monograph Series, 102*. Retrieved from https://bmcr.brynmawr.edu/1999/1999.04.21/

Burke, T. (2015). Manuscripts of the Ethiopic New Testament. In *The Textual History of the Greek New Testament: Changing Views in Contemporary Research* (pp. 437-452). SBL Press.

Burkett, D. (2002). *The Gospel According to John*. Oxford University Press.

Burkett, D. (2010). *An Introduction to the New Testament and the Origins of Christianity*. Cambridge University Press.

Burridge, R. A. (2004). *What Are the Gospels? A Comparison with Graeco-Roman Biography* (2nd ed.). Eerdmans.

Bynum, W. F., & Porter, R. (Eds.). (1993). *Companion Encyclopedia of the History of Medicine*. Routledge.

Cannon, J. W. (2018). Pathophysiology of hypovolemic shock. *Critical Care Clinics, 34*(1), 43-61.

Carrier, R. (2011–2012). Thallus and the Darkness at Christ's Death. *The Journal of Greco-Roman Christianity and Judaism*, 8, 185–191.

Charlesworth, J. H. (Ed.). (2006). *The Old Testament Pseudepigrapha, Volume 1: Apocalyptic Literature and Testaments*. Hendrickson Publishers.

Charlesworth, J. H. (Ed.). (2010). *The Old Testament Pseudepigrapha and the New Testament: Prolegomena for the Study of Christian Origins*. Cambridge University Press.

Chilton, B., & Neusner, J. (2005). *The zealots: Investigations into the Jewish freedom movement in the period from Herod I until 70 A.D.* Wipf and Stock Publishers.

Cohen, S. J. D. (2006). *From Maccabees to Mishnah* (2nd ed.). Westminster John Knox Press.

Cohn-Sherbok, D. (1996). *Messianic Expectation in Judaism: Its Historical Development*. Continuum.

Cohn-Sherbok, D. (2018). *Judaism: History, belief, and practice*. Routledge.

Collins, J. J. (2010). The zealots: The historical context of a Jewish resistance movement. In *Ancient Judaism and Christian Origins* (pp. 57-86). Brill.

Collins, R. (1995). Understanding Atonement: A New and Orthodox Theory. Grantham: Messiah College. Retrieved September 8, 2013.

Cook, J. G. (2014). *Crucifixion in the Mediterranean World*. Mohr Siebeck.

Cross, F. L. (Ed.). (2005). The Oxford dictionary of the Christian Church (p. 124, entry "Atonement"). Oxford University Press.

Crossan, J. D., & Reed, J. L. (2004). *Excavating Jesus: Beneath the Stones, Behind the Texts*. HarperCollins.

Cureton, W. (1858). *Remains of a Very Ancient Recension of the Four Gospels in Syriac, Hitherto Unknown in Europe*. London.

Dafni, A., Levy, S., & Lev, E. (2005). The ethnobotany of Christ's Thorn Jujube (Ziziphus spina-christi) in Israel. *Journal of Ethnobiology and Ethnomedicine, 1*(8). https://doi.org/10.1186/1746-4269-1-8

Defending Inerrancy. (n.d.). *Were the New Testament Manuscripts Copied Accurately?* Retrieved from https://www.defendinginerrancy.com/were-the-new-testament-manuscripts-copied-accurately/

Dibelius, M. (1927). The Structure and Literary Character of the Gospels. *The Harvard Theological Review, 20*(3), 151–170. Retrieved from http://www.jstor.org/stable/1507902

Droge, A. J. (2010). Roman Provincial Administration. In *The Oxford Handbook of Roman Studies*. Oxford University Press.

Dubin, A. E., & Patapoutian, A. (2010, November). Nociceptors: the sensors of the pain pathway. *Journal of Clinical Investigation, 120*(11), 3760-3772. doi: 10.1172/JCI42843

Edwards, W. D., Gabel, W. J., & Hosmer, F. E. (1986). On the Physical Death of Jesus Christ. *Journal of the American Medical Association, 255*(11), 1455-1463.

Encyclopædia Britannica. (2009). Encyclopædia Britannica Online: crucifixion. Retrieved December 19, 2009, from Britannica.com.

Encyclopedia Britannica. (n.d.). Liberation theology. In *Encyclopedia Britannica*. Retrieved October 14, 2021, from https://www.britannica.com/topic/liberation-theology

Eusebius of Caesarea. (1989). *The History of the Church: From Christ to Constantine* (G. A. Williamson, Trans.; A. Louth, Rev.). Penguin Classics.

Evans, C. A. (2012). *Jesus and His World: The Archaeological Evidence*. Westminster John Knox Press.

Evans, C. A. (2012). *The Gospels and Acts*. Baker Academic.

Evans, C. A. (2012). *The Historical Christ and the Jesus of Faith: The Incarnational Narrative as History*. Oxford University Press.

Evans, C.A. (2012). *The Reliability of the New Testament: Bart Ehrman and Daniel Wallace in Dialogue*. Fortress Press.

Faraoanu, E. (n.d.). Title of the Document. Retrieved from http://www.ejst.tuiasi.ro/Files/74/18_Faraoanu.pdf

Feldman, J. L., Del Negro, C. A., & Gray, P. A. (2013). Understanding the organization of respiratory rhythmogenesis: Insights from developmental neurobiology. *Frontiers in Cellular Neuroscience, 7*, Article 55.

Feldman, L. H. (2017). *Judaism and Hellenism Reconsidered*. Wipf and Stock Publishers.

Flavius Josephus. (c. 94 AD). *Antiquities of the Jews*. Book 18, Chapter 3.

Foster, P. (2007). *The Writings of the Apostolic Fathers*. T&T Clark.

Frederick, N. J. (2017). The Use of the Old Testament in the New Testament Gospels. In A. P. Schade, B. M. Hauglid, & K. Muhlestein (Eds.), *Prophets and Prophecies of the Old Testament* (pp. 123-160). Religious Studies Center; Deseret Book.

Gates, M. (2018). Crucifixion in the Ancient World: The Evidence. Biblical Archaeology Society. Retrieved from https://www.biblicalarchaeology.org/daily/ancient-cultures/daily-life-and-practice/crucifixion-in-the-ancient-world/

Goldstein, D. S. (2011). *Principles of Autonomic Medicine*. National Institutes of Health.

Goodman, M. (1992). *The Ruling Class of Judaea: The Origins of the Jewish Revolt against Rome, AD 66-70*. Cambridge University Press.

Goodman, M. (2008). *Rome and Jerusalem: The clash of ancient civilizations*. Vintage. (Note: This work is listed three times; if these are identical, only one entry is necessary unless there are different editions or forewords to distinguish them.)

Goodman, M. (2014). *The Roman World 44 BC–AD 180*. Routledge.

Grant, R. M. (2014). *Gods and the One God*. Westminster John Knox Press.

Gray, R. (2009). *A brief introduction to the New Testament* (2nd ed.). Westminster John Knox Press.

Gray, R. (2009). *The world of the New Testament: Cultural, social, and historical contexts*. Baker Academic.

Green, J. B. (1997). *The Gospel of Luke*. Eerdmans.

Guarino, M., Perna, B., Cesaro, A. E., Maritati, M., Spampinato, M. D., Contini, C., & De Giorgio, R. (2023). 2023 Update on Sepsis and Septic Shock in Adult Patients: Management in the Emergency Department. *Journal of Clinical Medicine, 12*(9), 3188. https://doi.org/10.3390/jcm12093188

Gutiérrez, G. (1988). *A Theology of Liberation: History, Politics, and Salvation*. Orbis Books.

Hall, J. E., & Guyton, A. C. (2020). *Guyton and Hall Textbook of Medical Physiology* (14th ed.). Elsevier.

Hall, J. E., & Guyton, A. C. (2020). *Guyton and Hall Textbook of Medical Physiology* (14th ed.). Elsevier.

Hankinson, R. J. (1998). *Cause and Explanation in Ancient Greek Thought*. Oxford University Press.

Hartog, P. (2015). *Polycarp and the New Testament: The Occasion, Rhetoric, Theme, and Unity of the Epistle to the Philippians and its Allusions to New Testament Literature*. Mohr Siebeck.

Hemer, C. J. (1990). *The Book of Acts in the Setting of Hellenistic History*. Edited by Conrad H. Gempf. J.C.B. Mohr (Paul Siebeck).

Hengel, M. (1974). *Judaism and Hellenism: Studies in Their Encounter in Palestine During the Early Hellenistic Period* (Vol. 1). Fortress Press.

Hengel, M. (1977). *Crucifixion in the ancient world and the folly of the message of the cross*. Fortress Press.

Hengel, M. (1977). *The Son of God: The Origin of Christology and the History of Jewish-Hellenistic Religion*. Fortress Press.

Hengel, M. (1989). *Judaism and Hellenism: Studies in Their Encounter in Palestine During the Early Hellenistic Period*. Fortress Press.

Hengel, M. (1989). *The Zealots: Investigations into the Jewish freedom movement in the period from Herod I until 70 A.D.* T&T Clark.

Heschel, S. (2009). *Crucifixion in the Ancient World and the Folly of the Message of the Cross*. Fortress Press.

Himmelfarb, M. (2005). *A kingdom of priests: Ancestry and merit in ancient Judaism*. University of Pennsylvania Press.

Horden, P. (2000). *The History of Medicine: A Beginner's Guide*. Oxford University Press.

Horsley, R. A. (1995). *Jesus and the spiral of violence: Popular Jewish resistance in Roman Palestine*. HarperCollins.

Horsley, R. A. (1997). *Bandits, prophets, and messiahs: Popular movements in the time of Jesus*. HarperSanFrancisco.

Horsley, R. A. (2001). *Bandits, Prophets, and Messiahs: Popular Movements at the Time of Jesus*. Trinity Press International.

Horsley, R. A. (Ed.). (1995). *Paul and politics: Ekklesia, Israel, imperium, interpretation*. Trinity Press International.

Houghton, H. A. G. (2016). *The Latin New Testament: A Guide to its Early History, Texts, and Manuscripts*. Oxford University Press.

Hunter, K. M. (1991). *Doctors' Stories: The Narrative Structure of Medical Knowledge*. Princeton University Press.

Hyvernat, E. (1914). *Coptic Versions of the Bible*. In *The Catholic Encyclopedia*. New York: The Encyclopedia Press. Retrieved from http://www.newadvent.org/cathen/16078c.htm

Jackson, R. (1988). *Doctors and Diseases in the Roman Empire*. University of Oklahoma Press.

Jefford, C. N. (Ed.). (1999). *The Apostolic Fathers: An Essential Guide*. Abingdon Press.

Jeffrey, D. L. (1992). *A Dictionary of Biblical Tradition in English Literature*. Wm. B. Eerdmans Publishing. ISBN 978-0-85244-224-1.

Johnson, L. T. (2010). *The New Testament: A very short introduction*. Oxford University Press.

Josephus, F. (1987). *The Jewish War* (G. Williamson, Trans.). Penguin Classics.

Josephus, F. (2004). *Antiquities of the Jews* (Penguin Classics ed.). Penguin Classics.

Josephus, F. (2016). *The Jewish War*. Princeton University Press.

Josephus, F. (75 AD). *The Jewish Wars*. Retrieved from https://www.gutenberg.org/files/2848/2848-h/2848-h.htm

Jouanna, J. (1999). *Hippocrates*. Johns Hopkins University Press.

Kahn, S. R., Lim, W., Dunn, A. S., et al. (2012). Prevention of VTE in Nonsurgical Patients: Antithrombotic Therapy and Prevention of Thrombosis, 9th ed: American College of Chest Physicians Evidence-Based Clinical Practice Guidelines. *Chest, 141*(2_suppl), e195S-e226S.

Karmy-Jones, R., Nathens, A., & Jurkovich, G. J. (2004). Management of traumatic lung injury: A Western Trauma Association Multicenter review. *Journal of Trauma and Acute Care Surgery, 57*(6), 1343-1349.

Keener, C. S. (2012). *The Gospel of Matthew: A Socio-Rhetorical Commentary*. Eerdmans.

Kellaway, J. (2003). *The History of Torture and Execution: From Early Civilization through Medieval Times to the Present*. Lyons Press.

King, H. (1998). *Health in Antiquity*. Routledge.

King, H. (2001). *Greek and Roman Medicine*. Bristol Classical Press.

Kulshrestha, P., Munshi, I., & Wait, R. (2004). Profile of chest trauma in a level I trauma center. *Journal of Trauma, 57*(3), 576-581.

Kumar, V., Abbas, A. K., & Aster, J. C. (2020). *Robbins and Cotran Pathologic Basis of Disease*. Elsevier Health Sciences.

Landén, N. X., Li, D., & Ståhle, M. (2016, October). Transition from inflammation to proliferation: a critical step during wound healing. *Cellular and Molecular Life Sciences, 73*(20), 3861-3885. doi: 10.1007/s00018-016-2268-0

Levine, L. I. (2005). *The ancient synagogue: The first thousand years* (2nd ed.). Yale University Press.

Levine, L. I. (2005). *The economic background to the Gospels*. Wipf and Stock Publishers.

Lewis, T., & Short, C. (n.d.). *A Latin Dictionary | vulgo*. Retrieved October 5, 2019, from http://www.perseus.tufts.edu

Li, J. (2019). *Evolution of Diseases in Historical Perspective*. Springer.

Lieu, J. (2018). *The Oxford Handbook of Early Christian Studies*. Oxford University Press.

Lucian of Samosata. (Mid-2nd century). *The Death of Peregrine*.

MacDonald, M. T. (2014). The Deutero-Pauline Letters in Contemporary Research. In *The Oxford Handbook of Pauline Studies*. New York: Oxford University Press.

MacMullen, R. (1984). *Christianizing the Roman Empire (A.D. 100-400)*. Yale University Press.

Madea, B. (2016). *Handbook of Forensic Medicine*. Wiley-Blackwell.
Markham, J. (2016). *Illness in Antiquity: A Cultural and Linguistic History*. Ashgate.
Martin, D. B. (2014). *New Testament history and literature*. Yale University Press.
Martin, D. B. (2014). *The Corinthian body*. Yale University Press.
Mason, S. (2016). *A history of the Jewish War, AD 66-74*. Cambridge University Press. (Note: Listed twice with slight variation, consider merging if the same work.)
Mason, S. N. (2018). The punitive power of the Roman Empire. *The Historian, 80*(1), 12-17. https://doi.org/10.1111/hisn.12685
Matthay, M. A., Zemans, R. L., Zimmerman, G. A., et al. (2019). Acute Respiratory Distress Syndrome. *Nature Reviews Disease Primers, 5*(1), 18.
Meier, J. P. (1991). *A Marginal Jew: Rethinking the Historical Jesus, Volume I: The Roots of the Problem and the Person*. Doubleday.
Melendez Rivera, J. G., & Anjum, F. (2023, April 27). Hypovolemia. In *StatPearls [Internet]*. StatPearls Publishing. Retrieved from https://www.ncbi.nlm.nih.gov/books/NBK565845/
Metzger, B. M., & Ehrman, B. D. (2005). *The Text of the New Testament: Its Transmission, Corruption, and Restoration*. Oxford University Press.
Metzger, B.M. (2005). *A Textual Commentary on the Greek New Testament*. Hendrickson Publishers.
Metzger, Bruce M. (2005). *The Text of the New Testament: Its Transmission, Corruption, and Restoration*. Oxford University Press.
Meyer, M. (2007). *The Nag Hammadi Scriptures: The International Edition*. HarperOne. ISBN 0-06-052378-6
Montefiore, S. S. (2011). *Jerusalem: The Biography*. Alfred A. Knopf.
Moore, E. E., Feliciano, D. V., & Mattox, K. L. (Eds.). (2017). *Trauma*. McGraw-Hill Education.
Moore, K. L., Dalley, A. F., & Agur, A. M. R. (2018). *Clinically Oriented Anatomy*. Wolters Kluwer.
Mora, S., et al. (2005). Physical activity and reduced risk of cardiovascular diseases: biological and epidemiological evidence. *The Lancet, 366*(9499), 1789-1799.
Murphy-O'Connor, J. (2008). *The Holy Land: An Oxford Archaeological Guide from Earliest Times to 1700* (5th ed.). Oxford University Press.
Myerburg, R. J., & Junttila, M. J. (2012). Sudden Cardiac Death Caused by Coronary Heart Disease. *Circulation, 125*(8), 1043-1052.
Naulty, M. (2012). The Physiology of Crucifixion. *Journal of the Royal Society of Medicine, 105*(4), 115–116.
Nestle, E., Nestle, E., Aland, B., & Aland, K. (Eds.). (2001). *Novum Testamentum Graece* (27th ed.). Stuttgart: Deutsche Bibelgesellschaft.
Netzer, E. (2001). *The architecture of Herod, the Great Builder*. Mohr Siebeck. (Note: Listed twice with slight variation, consider merging if the same work.)
Neusner, J. (1973). *A History of the Jews in Babylonia: The Age of Shapur II*. Brill.
Neusner, J. (1975). *Judaism in the Beginning of Christianity*. Fortress Press.
Neusner, J. (1988). *Judaism and Christianity in the Age of Constantine: History, Messiah, Israel, and the Initial Confrontation*. University of Chicago Press.
Niewöhner, P. (2016). The Archaeology of Crucifixion. *Near Eastern Archaeology, 79*(4), 214-219. https://doi.org/10.5615/neareastarch.79.4.0214
Nikolich-Žugich, J., et al. (2008). Ageing and life-long maintenance of T-cell subsets in the face of latent persistent infections. *Nature Reviews Immunology, 8*(7), 512-522.
Nutton, V. (2013). *Ancient Medicine*. Routledge.
Pagels, E. (1979). *The Gnostic Gospels*. Random House.
Parker, D. C. (2008). Textual criticism. In *An Introduction to the New Testament Manuscripts and their Texts*. Cambridge University Press. DOI: 10.1017/CBO9780511619922.006.
Pearn, J. (2014). Pathophysiology of drowning. *Australian Family Physician, 43*(6), 370-373.
Phang, S. E. (2008). *Roman Military Service: Ideologies of Discipline in the Late Republic and Early Principate*. Cambridge University Press.
Philo. (1993). Embassy to Gaius. In F. H. Colson & G. H. Whitaker (Eds.), *Philo* (Vol. 10, pp. 71-141). Harvard University Press.

Pliny the Younger. (112 AD). *Letters* (Vol. 1) (B. Radice, Trans.). Harvard University Press.

Plutarch. (1914). *Plutarch's Lives* (B. Perrin, Trans.). Harvard University Press.

Pormann, P. E., & Savage-Smith, E. (2007). *Medieval Islamic Medicine*. Edinburgh University Press.

Porter, R. (1997). *The Greatest Benefit to Mankind: A Medical History of Humanity*. W.W. Norton & Company.

Porter, S. E. (1997). *The Language of the New Testament: Classic Essays*. Sheffield Academic Press.

Purves, D., Augustine, G. J., Fitzpatrick, D., et al. (2020). *Neuroscience*. Sinauer Associates.

Rasmussen, C. G. (2013). *Zondervan Essential Atlas of the Bible*. Zondervan.

Reed, J. L. (2002). *Archaeology and the Galilean Jesus: A Re-examination of the Evidence*. Trinity Press International.

Religious Studies Center. (n.d.). *New Testament Manuscripts, Textual Families, and Variants*. Retrieved from https://rsc.byu.edu/new-testament-manuscripts-textual-families-and-variants

Retief, F. P., & Cilliers, L. (2003). The history and pathology of crucifixion. *South African Medical Journal, 93*(12), 938-941. PMID: 14750495.

Rhoads, D. M. (2011). *Israel in revolution* (2nd ed.). Baylor University Press. (Note: Listed twice with slight variation, consider merging if the same work.)

Roberts, Donaldson & Coxe. (1896). In "Contra Celsum", Book II, chapter 14,23,59 (p. 441). Volume IV.

Roberts, M. D. (2010). *Jesus and Aramaic in the Gospels*. Retrieved from https://www.beliefnet.com/columnists/markdroberts/2010/07/jesus-and-aramaic-in-the-gospels.html

Robinson, J. A. T. (1976). *Redating the New Testament*. Westminster Press.

Robinson, J. M. (Ed.). (1990). *The Nag Hammadi Library in English* (3rd ed.). HarperSanFrancisco.

Sanders, E. P. (1993). *Jesus and Judaism*. Fortress Press.

Sanders, E. P. (1993). *Judaism: Practice and belief, 63 BCE–66 CE*. SCM Press.

Saukko, P., & Knight, B. (2016). *Knight's Forensic Pathology*. CRC Press.

Scarborough, J. (1992). *Medical and Biological Terminologies: Classical Origins*. University of Oklahoma Press.

Schäfer, P. (2003; 2014). *The history of the Jews in the Greco-Roman world: The Jews of Palestine from Alexander the Great to the Arab conquest*. Routledge. (Note: Listed three times, consider merging if the same work.)

Schwartz, D. R. (2018). *The First Jewish Revolt: Archaeology, History, and Ideology*. Routledge.

Segal, A. F. (2018). *The concise dictionary of Judaism*. Rowman & Littlefield.

Siraisi, N. G. (1990). *Medieval and Early Renaissance Medicine: An Introduction to Knowledge and Practice*. University of Chicago Press.

Smith, J. (Year). Crucifixion: A Historical Analysis of the Cause of Death. *Journal of Ancient Punishments, 25*(2), 145-162. (Publication year not provided, indicated as "Year".)

Spodick, D. H. (2003). Acute cardiac tamponade. *New England Journal of Medicine, 349*(7), 684-690.

Stern, S. J. (2013). *Jewish identity in the Greco-Roman world*. University of Pennsylvania Press.

Stone, M. E. (2014). Armenian Versions. In *The Textual History of the Greek New Testament: Changing Views in Contemporary Research* (pp. 453-468). SBL Press.

Swearingen, P. L. (2016). *All-In-One Nursing Care Planning Resource*. Elsevier.

Tacitus. (1934). *The Annals* (J. Jackson, Trans.). Loeb Classical Library.

Tacitus. Annals IV. 72, 3. (No publication year provided for a precise APA citation.)

Tapson, V. F. (2008). Acute Pulmonary Embolism. *New England Journal of Medicine, 358*(10), 1037-1052.

The Center for the Study of New Testament Manuscripts. (n.d.). *Manuscripts 101: What is a textual variant?* Retrieved from https://www.csntm.org/manuscripts-101-what-is-a-textual-variant

Thompson, C., & Venables, R. (2017). *Cultural Contexts of Health: The Use of Narrative in Medicine*. Brill.

Thompson, J. A. (1997). *The Bible and archaeology*. Wm. B. Eerdmans Publishing.

Timmers, I., Quaedflieg, C. W. E. M., Hsu, C., Heathcote, L. C., Rovnaghi, C. R., & Simons, L. E. (2019, December). The interaction between stress and chronic pain through the lens of threat learning. *Neuroscience and Biobehavioral Reviews, 107*, 641-655. doi: 10.1016/j.neubiorev.2019.10.007

Tristram, H. B. (2007). *The Physical Geography, Geology, and Meteorology of the Holyland* (ISBN 1593334826). Page 11.

Turner, R. (2009, October 1). An analysis of the Pre-Pauline creed in 1 Corinthians 15:1-11. Retrieved from https://carm.org/about-pre-pauline-creed-in-1-corinthians-15

University of Pennsylvania Museum of Archaeology and Anthropology. (n.d.). *Saint Mathew Gospel*. Penn Museum. Retrieved August 18, 2017, from https://www.penn.museum

VanderKam, J. C. (2010). *An introduction to early Judaism*. Eerdmans.

VanderKam, J. C., & Flint, P. W. (2012). *The meaning of the Dead Sea Scrolls: Their significance for understanding the Bible, Judaism, Jesus, and Christianity*. HarperCollins.

Vermes, G. (1997). *The Complete Dead Sea Scrolls in English*. Penguin Books.

Vermes, G. (2014). *The true history of the first Easter*. Bloomsbury Publishing.

Wachsmann, S. (1995). *The Sea of Galilee Boat: An Extraordinary 2000 Year Old Discovery*. Plenum Press.

Wallace, Daniel B. (2006). The Gospel According to Bart: A Review Article of Misquoting Jesus by Bart Ehrman. *Journal of the Evangelical Theological Society, 49*(2), 327-349.

Ware, L. B., & Matthay, M. A. (2000). The Acute Respiratory Distress Syndrome. *New England Journal of Medicine, 342*(18), 1334-1349.

Weaver, J. Denny. (2001). *The Nonviolent Atonement*. Wm. B. Eerdmans Publishing.

West, J. B. (2012). *Respiratory Physiology: The Essentials*. Wolters Kluwer Health.

Wettstein, J. J. (1751). *Novum Testamentum Graecum editionis receptae cum lectionibus variantibus codicum manuscripts* (pp. 8–41). Amsterdam: Ex Officina Dommeriana.

Wijdicks, E. F. M. (2001). Determining brain death in adults. *Neurology, 56*(10), 1005-1015.

Wijdicks, E. F. M. (2002). The diagnosis of brain death. *New England Journal of Medicine, 344*(16), 1215-1221.

Williams, M. (2021). Wound infections: an overview. *British Journal of Community Nursing, 26*(Sup6), S22-S25. https://doi.org/10.12968/bjcn.2021.26.Sup6.S22

Wittstein, I. S., Thiemann, D. R., Lima, J. A., et al. (2005). Neurohumoral Features of Myocardial Stunning Due to Sudden Emotional Stress. *New England Journal of Medicine, 352*(6), 539-548.

Wroe, A. (1999, April 3). Historical notes: Pontius Pilate: a name set in stone. *The Independent*.

Zhou, S. F., et al. (2009). Polymorphism of human cytochrome P450 enzymes and its clinical impact. *Drug Metabolism Reviews, 41*(2), 89-295.

Zias, J., & Sekeles, E. (1985). The Crucified Man from Giv'at ha-Mivtar: A Reappraisal. *Israel Exploration Journal, 35*(1), 22-27.

Zugibe, F. T. (2005). *The Crucifixion of Jesus: A Forensic Inquiry*. M. Evans and Company, Inc.

INDEX

A

abuse 112, 113, 117, 118, 130
acidosis 132, 134, 142, 143, 147, 162, 173, 175, 176, 198, 202
anachronism 193, 194
ancient narratives 195
ancient works 81
anemia 124
anguish 9, 30, 32, 117, 120, 121, 127, 129, 136
Antonia Fortress 113, 115
Apostolic Fathers 98, 99, 221, 222
Aramaic. 31, 88, 89, 98, 99, 219, 224
archaeological findings 3, 5, 8, 12, 29, 72, 86, 94, 96, 97, 114, 181
archaeology 41, 94, 97, 225
arteries 133, 144, 158, 203
Asphyxiation. 44, 143, 174, 198, 200, 202
atonement 6, 7, 8, 12, 18, 34, 39, 48, 49, 51, 58, 59, 63, 64, 65, 66, 67, 68, 69, 70
attestations 94, 217
authenticity37, 41, 42, 72, 86, 87, 88, 89, 91, 92, 99, 100, 102

B

bacteria 129
Barabbas 112
Bethsaida 95
betrayal 30, 31, 62, 110, 111, 121, 136
blood cells 124, 126, 139
blood pressure ... 125, 133, 134, 135, 141, 147, 151, 152, 167, 174, 198, 199, 201
blood vessels 125, 126, 129, 133, 142, 149, 152, 167, 173, 174
bodily functions .. 128, 135, 155, 156, 159, 165, 173
breathing difficulties 132, 133

C

Caesarea Maritima 16, 94, 96
Caiaphas 30, 31, 85, 111, 112
Capernaum 95, 106
cardiac arrest .. 9, 133, 146, 147, 173, 198, 211
cardiac rupture 140, 141, 145, 202, 205, 207, 208, 209, 210
Chorazin 95
Christ 52, 54, 67, 68, 86, 98, 103, 115, 116, 128, 207, 220
Christian communities .. 6, 33, 54, 58, 60, 74, 75, 76
Christian community7, 64, 89, 90, 91, 98, 99, 100
Christian ethics 60, 99
Christian practices 7, 98
Christian thought 6, 7, 56, 63, 64, 90, 91, 92, 99, 100, 103
Christian writings 80, 99, 101, 117
Christus Victor 7, 64, 65, 66
Church 54, 58, 60, 61, 63, 64, 65, 76, 77, 78, 79, 80, 98, 99, 103, 113, 116, 207, 220
Church Fathers 64, 65, 78, 79, 80
circumstances ... 26, 39, 99, 118, 143, 150, 151, 161, 162, 163, 167, 169, 171, 173, 174, 175, 204, 213, 217
clinical. 144, 146, 152, 153, 158, 159, 164, 165, 166, 168, 169, 170, 182, 186, 188, 194, 195, 204, 208, 225
Codex Alexandrinus 73, 74
Codex Sinaiticus 73, 78
Codex Vaticanus 73, 78
collapse 135, 141, 142, 145, 151, 152, 173, 174, 175, 176
concerns .. 57, 93, 191, 192, 193, 194
condemnation 31, 111, 112, 115, 117
confusion 124, 132, 173, 198
consciousness 55, 131, 132, 167, 168
crimes 32, 47, 87, 139, 201
criminals 16, 29, 84, 86, 87, 117, 197
critics .. 57, 64, 74, 75, 79, 80, 81, 82,

Index

90, 91, 105
crossbeam 6, 32, 43, 44, 45, 49, 114, 118, 119, 120, 121
crown 55, 112, 113, 128, 129, 130, 131, 136, 142
Crown of Thorns 128, 142

D

Dead Sea Scrolls 17, 23, 97, 101, 225
deficiency 124
diagnoses ... 140, 177, 178, 180, 181, 185, 186, 190, 203, 213
diagnostic 11, 145, 164, 165, 168, 169, 177, 179, 180, 181, 182, 184, 186, 187, 189, 195
Didache 98
discomfort 127
distress ...30, 44, 111, 128, 132, 133, 135, 136, 143, 145, 150, 151, 152, 161, 162, 163, 173, 176, 200, 217

E

early Christians 19, 55, 76, 91, 93, 94, 99
eclipse 34, 115
electrolyte imbalances 147
emotional toll 110
Essenes 4, 15, 17, 18, 22, 97
ethical concepts 107
Eucharist 7, 61, 62, 99
Eusebius of Caesarea 98, 220
excruciating ..33, 40, 44, 45, 47, 118, 127, 132, 138, 142, 171, 172, 197, 200, 202
executioners .6, 35, 46, 84, 130, 137, 138, 139, 204, 216, 217

G

Great Jewish Revolt 25

H

heart muscle 140, 142, 146, 147
heart rate 124, 125, 128, 133, 142, 144, 145, 147, 151, 174, 198, 199, 201
Hellenism 19, 21, 22, 26, 101, 221

Hemopericardium 140
Herod Antipas 112
high priest 26, 30, 31, 85, 111
High Priest 18, 19, 27
historical documents 71, 98, 104
historical event 13, 48, 52, 59, 61, 85, 217
historical records 5, 8, 40, 42, 51, 71, 72, 84, 86, 106, 110
Holy Sepulchre 113, 116
humility 11, 62, 130
hypercapnia 132, 143, 159, 162, 173, 176, 202
Hypotension 125
hypothesis .. 200, 202, 205, 206, 207, 208, 210
Hypovolemia 125, 223

I

identity 2, 4, 7, 18, 22, 23, 25, 34, 38, 39, 54, 61, 62, 111, 129, 219, 224
illness 145, 182, 183, 187
imbalances .. 147, 173, 178, 186, 189
immune system ... 124, 128, 135, 188
infection 124, 126, 128, 129, 135, 140, 145, 146
inflammation 126, 127, 140, 145, 150, 222
inflammatory 126, 145, 146, 178
inquiry 7, 11, 51, 169, 177, 185, 197, 209, 217
instability 20, 145
interpreting 11, 49, 75, 90, 91, 93, 104, 182, 183, 184, 190, 193, 204, 213
intervention 19, 21, 23, 27, 120, 147, 151, 153, 156, 172, 173, 217
isolation.. 30, 65, 111, 130, 136, 164, 168

J

Jericho 106
Jewish community 20, 24, 86, 99
Jewish feasts 97
Jewish law 17, 23, 26, 27, 28, 111, 117
Jewish practices 97, 100
Jewish rituals 97, 98
Jewish society . 4, 15, 17, 24, 26, 100
Jewish worship 18, 97

John Calvin 63
Jordan River................................ 106
Josephus.... 8, 17, 21, 22, 40, 41, 93, 94, 102, 221, 222
Judas Iscariot................ 30, 111, 136

L

lacerations................... 124, 141, 201
languages 33, 34, 72, 75, 80, 183, 193
Last Supper 8, 30, 68, 110, 111, 120, 209
life-threatening ... 124, 125, 129, 130, 144, 151, 209
love .7, 28, 32, 33, 35, 48, 49, 50, 52, 54, 55, 56, 57, 59, 60, 61, 62, 65, 67

M

manuscript 71, 72, 73, 74, 75, 76, 78, 79, 80, 82, 100
medical analysis. 3, 11, 12, 132, 182, 197, 203, 207, 208, 211
medical condition 110, 124, 133, 146, 185
medical conditions 124, 125, 182, 184, 190, 195
medical documentation 110, 191
medical language......................... 192
medical perspective ... 4, 8, 123, 200, 204, 205, 211
medical practices . 11, 159, 170, 179, 184, 185, 186, 187, 188, 191, 192, 193, 194
medical records................... 105, 123
metabolism. 134, 142, 164, 173, 188, 189
Mishnah 20, 22, 101, 220
morality 55, 56, 60, 107
Mount of Olives 30, 106, 111
myocardial damage..................... 147

N

nailed 32, 33, 43, 44, 45, 87, 144, 198, 200, 201
nailing .. 6, 8, 32, 33, 45, 49, 87, 133, 134, 142, 144, 145, 199, 206, 211
nails ..32, 33, 45, 131, 132, 133, 171, 172, 198, 204

Nazareth......3, 13, 18, 19, 20, 33, 53, 71, 95, 96, 104, 106
nervous system .. 127, 129, 138, 155, 156, 157, 158, 166, 168

O

organ failure 125, 134, 135, 141, 142, 152, 162, 175, 176, 198, 199, 200, 201
organ function....................... 134, 142
oxygen levels....... 128, 132, 143, 145, 146, 152, 161, 163, 173, 174, 175
oxygenation 125, 128, 144, 153, 164, 175

P

pain receptors...................... 126, 128
Pauline Epistles........................ 91, 92
perfusion...................... 133, 152, 198
pericardial..... 10, 140, 141, 149, 150, 151, 153, 154, 200, 204, 208
Peter..... 30, 54, 61, 65, 88, 111, 117, 136
Pharisees ... 4, 15, 17, 18, 19, 22, 27, 86, 97
Phlegon 103
physical agony........ 35, 129, 130, 133
physical assault 127, 131
physical condition 110, 118, 119, 123, 128, 131
physical stimulus 126
physically excruciating................. 113
physiological responses 140, 149, 154, 188, 203, 212
physiological stress 135, 146, 203, 205, 206, 211
physiological stresses. 132, 134, 145, 199, 212
Pliny the Younger 103, 224
Plutarch 40, 81, 224
Polybius... 81
Polycarp of Smyrna 98
Pontius Pilate.. 15, 16, 31, 38, 85, 86, 93, 94, 96, 97, 102, 112, 113, 115, 123, 225
Pool of Siloam 94
possibility...... 1, 10, 53, 67, 105, 114, 152, 177, 207, 208, 217
post-mortem . 10, 140, 157, 158, 169, 170, 202, 208

Index

prayer 18, 30, 62, 109, 110, 111, 120
prophecies 19, 39, 53, 69, 89, 90, 111
Prostaglandins 126
Protestant Reformation 61
Psalms 89
pulmonary 144, 145, 203, 211

R

Rabbi Joshua ben Perachiah 104
red blood cells 124, 125
religious texts 80, 90, 91, 94, 98, 102, 104, 115, 199
respiratory alkalosis 132
respiratory failure 133, 139, 143, 146, 152, 153, 176, 198, 199, 200, 202
risk factors 145
Roman execution methods 86, 87, 114
Roman law 4, 5, 15, 16, 22, 31, 46, 197
Roman practice .. 10, 29, 41, 87, 112, 200
Roman rule . 5, 15, 16, 17, 19, 20, 21, 22, 23, 24, 27, 28, 29, 41, 112
Roman soldier 10, 36, 84, 200
Roman whip 123

S

Sadducees .. 4, 15, 17, 18, 23, 26, 27, 86, 97
salvation ... 4, 6, 7, 35, 36, 39, 48, 49, 51, 53, 56, 57, 62, 63, 66, 67, 68, 69, 91, 121
Sanhedrin . 16, 19, 25, 26, 27, 31, 38, 39, 85, 86, 103, 104, 111, 112
Satan 22, 64
Simon of Cyrene 32, 84, 87, 116, 120
sins 18, 32, 33, 36, 44, 48, 52, 53, 63, 66, 68, 69, 90, 136
skepticism 9, 107
skeptics 4, 107, 110, 216
slaves 25, 40, 54, 197

social justice 4, 21, 52, 56, 57, 58, 60
spear 10, 36, 139, 149, 150, 151, 152, 153, 154, 165, 166, 169, 199, 200, 202, 208
speculations 117
stamina 105, 106, 107, 108, 109, 110, 120
synagogues 17, 18, 107
Synoptic 120

T

Tacitus 8, 40, 43, 81, 92, 93, 94, 102, 224
Talmud 18, 102, 103, 104
Talmuds 101
Temple . 4, 15, 17, 18, 19, 22, 23, 24, 25, 26, 27, 97, 98, 100
testimonies 31, 85, 88, 111
textual analysis 75, 81, 100, 206, 207
textual criticism 72, 73, 74, 75, 76, 78, 79, 80, 81
Textual Criticism 74, 79, 81, 219
Thallus 103, 220
Thucydides 81
Tiberias 95
Torah 15, 17, 18, 19, 26, 97
torture 9, 84, 121, 129, 137, 177, 197, 203, 217
traumatic beatings 125
traumatic event 151
trials 8, 31, 111, 117, 120
truth 12, 83, 163, 217

V

versions 73, 75, 76, 77, 79
Via Dolorosa ... 8, 113, 115, 116, 117, 118, 119, 121, 126, 136
vital organs. 125, 128, 133, 142, 152, 162, 168, 174, 198, 200

X

Xenophon 82